HOW TO USE A PLASMA CUTTER

The Essential Handbook for Metal Artists and Fabricators

Fitzpatrick J. Thompkins

Copyright © 2024 by **Fitzpatrick J. Thompkins**

All rights reserved

No part of this publication may be reproduced, stored in a retrieval system, or transmitted, in any form or by any means, electronic, mechanical, photocopying, recording, or otherwise, without the prior written permission of the author.

The information in this ebook is true and complete to the best of our knowledge. All recommendation are made without guarantee on the part of author or publisher. The author and publisher disclaim any liability in connection with the use of this information.

Table of Contents

Introduction   5
   Overview of Plasma Cutting   7
   Benefits of Using a Plasma Cutter   8
   Safety Precautions and Protective Gear   10
   Chapter: 1 Understanding Plasma Cutters   12
   The Science Behind Plasma Cutting   12
   Components of a Plasma Cutter   14
   Types of Plasma Cutters   16
      Manual Plasma Cutters   16
      Automated/Computer Numerical Control (CNC) Plasma Cutters   16
      Understanding Plasma Cutters   17

Chapter: 2 Setting Up Your Plasma Cutter   18
   Initial Setup and Safety Checks   18
   Assembling the Plasma Cutter   20
   Connecting to a Power Source   22
   Air Supply Connection and Regulation   24

Chapter:3 Preparing to Cut   26
   Material Preparation and Placement   26
   Setting the Correct Amperage and Air Pressure   28
   Positioning and Securing the Workpiece   30
   Marking and Template Usage   32

Chapter: 4 Operating the Plasma Cutter   34

Safety Precautions Before Cutting	34
Igniting the Plasma Torch	36
Techniques for Straight Cuts	38
Techniques for Curved Cuts and Holes	40
Managing Speed and Movement	42
Troubleshooting Common Issues	44
Chapter: 5 Advanced Techniques	**46**
Piercing Thick Materials	46
Beveling Edges	48
Cutting Stacked Layers	50
Utilizing CNC Plasma Cutting for Precision Work	52
Chapter: 6 Maintenance and Troubleshooting	**54**
Routine Maintenance Checks	54
Replacing Consumables	56
Troubleshooting Cutting Problems	58
Dealing with Electrical Issues	60
Chapter: 7 Applications of Plasma Cutting	**62**
Industrial Applications	62
Artistic and Decorative Uses	64
Repair and Maintenance Work	66
Chapter: 8 Health and Safety Considerations	**68**
Ventilation and Fume Extraction	68
Noise and Ear Protection	70
Eye Protection and UV Radiation	72
Proper Handling and Ergonomics	74

Chapter: 9 Comparing Plasma Cutting with Other Cutting

Methods   **76**

  Plasma vs. Oxy-Fuel Cutting   76

    When comparing the two methods, several key differences emerge: 76

  Plasma vs. Laser Cutting   78

    Efficiency and Material Thickness   78

    Cut Quality and Precision   78

    Material Compatibility   78

    Operational Cost and Maintenance   79

    Safety and Environmental Considerations   79

  Advantages and Limitations   79

Conclusion   **81**

Introduction

In the heart of a bustling workshop filled with the hum of machinery and the sharp scent of metal, Sam, a seasoned metal artist, stood contemplating a slab of steel. His latest project was ambitious, a large-scale metal sculpture designed to captivate and inspire. However, Sam faced a challenge: his traditional tools could not achieve the precision and efficiency required for this intricate work. That's when he stumbled upon "HOW TO USE A PLASMA CUTTER," a book that promised to unlock the full potential of plasma cutting technology.

The book, written by renowned metalwork expert Alex Rivera, was a comprehensive guide that combined theoretical knowledge with practical applications, tailored for both beginners and seasoned professionals. Skeptical yet intrigued, Sam decided to purchase the book, hoping it would be the solution he was searching for.

As Sam turned the pages, he was introduced to the world of plasma cutting, a technique that uses a jet of superheated plasma to cut through metal with incredible precision and speed. The book meticulously guided him through every aspect of using a plasma cutter, from setting up the equipment and safety precautions to mastering advanced cutting techniques. It was as if Alex Rivera had condensed years of professional experience into a single, accessible guide.

One of the book's highlights was its emphasis on safety and best practices, teaching Sam how to minimize risks while maximizing the tool's effectiveness. Furthermore, the detailed illustrations and step-by-step instructions demystified the process, making complex techniques seem attainable. Sam was particularly fascinated by the section on artistic and decorative uses of plasma cutting, which expanded his vision for what was possible in his own work.

As days turned into weeks, Sam applied the knowledge he gained from the book to his project. The difference was night and day. His cuts were cleaner, his work pace quicker, and the details in his sculpture more refined than ever before. The plasma cutter, once a daunting piece of equipment, had become an extension of his creative will, thanks to "HOW TO USE A PLASMA CUTTER."

The sculpture, when completed, was a testament not only to Sam's skill as an artist but also to the transformative power of the right knowledge. It stood proudly in the local art gallery, drawing admiration from visitors who marveled at its intricate details and the smoothness of its curves, all made possible by the precision of plasma cutting.

Reflecting on his journey, Sam realized that purchasing the book was one of the best decisions he had made for his craft. "HOW TO USE A PLASMA CUTTER" had not only taught him a new skill but had also inspired him to push the boundaries of his art. It was a reminder that in the right hands, with the right knowledge, the possibilities of creation were limitless.

To anyone standing at the edge of discovery, wondering whether to dive into the world of plasma cutting, Sam's story serves as a beacon. "HOW TO USE A PLASMA CUTTER" is more than just a book; it's a key to unlocking potential, a guide for those ready to transform their work and embrace the future of metalworking.

Overview of Plasma Cutting

Plasma cutting stands out as a pivotal technology in the metalworking industry, offering unparalleled precision, efficiency, and versatility in cutting through various types of metal. This cutting-edge method employs a plasma torch to cut through electrically conductive materials, including steel, stainless steel, aluminum, brass, and copper, making it an indispensable tool for fabricators, sculptors, and anyone involved in metalwork. At its core, plasma cutting operates by sending an electric arc through a gas that is passed through a constricted opening. The gas can be air, nitrogen, argon, or oxygen. This process increases the temperature of the gas to the point where it enters a fourth state of matter known as plasma. This highly energized, ionized gas then becomes capable of conducting electricity, allowing the plasma cutter to complete its electrical circuit with the metal being cut. As the plasma jet hits the workpiece, it melts the metal, and a high-velocity stream of ionized gas mechanically blows the molten metal away, creating a clean cut with remarkable speed.

The versatility of plasma cutting is further enhanced by its ability to perform well in both hand-held and automated settings. Hand-held plasma cutters are ideal for portable applications and on-site jobs, where flexibility and ease of use are paramount. On the other hand, automated or CNC (Computer Numerical Control) plasma cutting systems offer unmatched precision and are typically used in industrial settings where complex, repetitive patterns and high production volumes are the norm.

One of the key advantages of plasma cutting over traditional metal cutting techniques is its speed and efficiency. Plasma cutters can cut metal much faster than oxy-fuel cutters and are capable of making precise cuts with minimal heat input, reducing the heat-affected zone and preserving the integrity of the material being cut. This results in less warping, especially when working with thin sheets of metal. Furthermore, plasma cutting does not rely on preheating the metal, making it possible to start cutting almost immediately after the machine is powered on, saving valuable time in fast-paced work environments.

Safety is another critical aspect of plasma cutting, as the process involves high temperatures, bright light, and ultraviolet radiation. Proper safety gear, including eye protection, gloves, and protective clothing, is essential to prevent burns, injuries, and exposure to harmful light emissions. Additionally, adequate ventilation is necessary to remove potentially hazardous fumes and gases produced during the cutting process.

In summary, plasma cutting is a powerful, efficient, and versatile method of cutting metal that has revolutionized the metalworking industry. Its ability to quickly and accurately cut through various metals with minimal waste and deformation has made it an essential tool in a wide range of applications, from industrial manufacturing to artistic metal sculpture. Whether utilized in hand-held devices or integrated into complex CNC systems, plasma cutting technology offers a blend of speed, precision, and flexibility that traditional metal cutting methods simply cannot match.

Benefits of Using a Plasma Cutter

Plasma cutting technology offers a myriad of benefits for individuals in metalworking fields, providing a versatile and efficient method for cutting through various types of metal. The advent of plasma cutters has revolutionized the way metal is cut, offering precision and speed that traditional cutting methods cannot match. One of the most significant advantages of using a plasma cutter is its ability to cut through different metal types and thicknesses with ease, ranging from thin sheets to thick plates. This versatility makes it an invaluable tool for a wide array of applications, from industrial manufacturing to artistic metalwork.

The precision of plasma cutters is unparalleled, producing clean and smooth cuts that require minimal finishing. This high level of accuracy not only saves time in the cutting process but also reduces the amount of waste produced, making it a cost-effective solution for metal cutting. The speed at which plasma cutters operate is another noteworthy benefit. They can cut metal much faster than oxy-fuel cutters and saws, significantly increasing productivity. This speed, combined with precision, allows for the completion of projects in a fraction of the time it would take using traditional cutting methods.

Plasma cutters are also renowned for their ease of use. With proper guidance, such as that provided in "HOW TO USE A PLASMA CUTTER," operators can quickly learn how to use these tools effectively, regardless of their previous experience with metal cutting. The learning curve is relatively short, and with the right information, users can achieve professional-quality cuts in no time. This accessibility opens up the world of metal cutting to hobbyists and professionals alike, allowing for more creative and innovative projects.

Another benefit of plasma cutting is its portability. Many plasma cutters are designed to be lightweight and compact, making them easy to transport between job sites or move around within a workshop. This portability, combined with the minimal setup required, makes plasma cutters an excellent option for on-site work and applications where mobility is crucial.

Safety is a paramount concern in any form of metalworking, and plasma cutters offer features that make them safer to use than many traditional cutting methods. The use of a non-flammable gas and the absence of open flames reduce the risk of accidents and fires. Additionally, with proper safety gear and adherence to safety guidelines, the risk of personal injury is minimized, making plasma cutting a safer alternative to oxy-fuel cutting and other techniques that involve open flames or hazardous materials.

The environmental impact of cutting methods is an increasing concern, and plasma cutters offer a more environmentally friendly option. They produce less waste and fewer toxic gases compared to traditional cutting methods, contributing to a healthier work environment and reducing the ecological footprint of metalworking projects.

In conclusion, the benefits of using a plasma cutter are substantial, offering a blend of efficiency, precision, safety, and versatility that traditional metal cutting methods cannot match. These advantages, coupled with the ease of use and environmental benefits, make plasma cutting a preferred choice for a wide range of metalworking applications. Through comprehensive guides like "HOW TO USE A PLASMA CUTTER," users can unlock the full potential of plasma cutting technology, enhancing their work quality and productivity while ensuring safety and minimizing environmental impact.

Safety Precautions and Protective Gear

When engaging in plasma cutting, a process that melds the precision of modern technology with the raw power of electrically ionized gas to cut through metal, the importance of safety precautions and the correct use of protective gear cannot be overstressed. This cutting-edge method offers significant advantages in terms of speed and precision but also presents unique hazards that require a well-informed approach to ensure personal safety and the safety of those in the immediate vicinity.

The cornerstone of safety in plasma cutting involves understanding the risks associated with the process. These risks include exposure to very bright light and ultraviolet (UV) radiation, which can harm the eyes and skin; the potential for burns from the hot metal or the torch tip; the danger of electric shock from the electrical components of the plasma cutter; and the inhalation of fumes and dust generated during cutting.

Protective gear is essential to mitigate these risks. A high-quality auto-darkening welding helmet is crucial, as it protects the eyes from the intense light and UV radiation emitted during the cutting process. This helmet should meet national safety standards and be adjustable to change the level of darkness based on the brightness of the plasma arc. Similarly, safety glasses with side shields or goggles should be worn under the helmet to provide additional protection for the eyes, especially from flying particles.

The skin should be protected from the high temperatures and UV radiation. This is achieved by wearing flame-resistant clothing that covers all exposed skin. Gloves made from leather or another flame-resistant material are necessary to prevent burns and cuts. These gloves should provide a balance between protection and dexterity to safely handle the equipment and workpiece.

Hearing protection, such as earplugs or earmuffs, guards against the loud noise produced by the plasma cutter, which can be damaging over time. Additionally, wearing leather boots or shoes can protect the feet from hot metal and sparks, while a leather apron can offer extra protection for the front of the body.

Ventilation is a critical aspect of safety that often requires more than just personal protective gear. Working in a well-ventilated area is essential to disperse the harmful fumes and dust generated during the cutting process. In some cases, it may be necessary to use a fume extraction system or respirator, especially in enclosed spaces or when working with materials that produce particularly toxic fumes.

Electrical safety is another key concern. Ensuring that the plasma cutter and all associated equipment are properly grounded can prevent electrical shock. It's also important to keep the work area dry and to wear insulated boots or stand on a dry, insulating mat when using the equipment.

In essence, the fusion of meticulous attention to safety protocols and the diligent use of protective gear forms the backbone of effective plasma cutting. This approach not only safeguards the well-being of the operator but also enhances the precision and quality of the work produced. By adhering to these safety measures, individuals can harness the full potential of plasma cutting technology, transforming raw metal into works of precision and artistry, all while maintaining the highest standards of safety and care.

Chapter: 1 Understanding Plasma Cutters

The Science Behind Plasma Cutting

Plasma cutting stands as a testament to human ingenuity, harnessing the fourth state of matter—plasma—to slice through metal with precision and efficiency. At its core, the science behind plasma cutting is a sophisticated blend of physics and technology, tailored to manipulate materials at the molecular level.

The journey begins with understanding what plasma is. In simple terms, plasma is ionized gas, a state achieved when a gas is subjected to extremely high temperatures or electrical energy, causing the gas molecules to become excited and ionized. This ionization process breaks the gas molecules into positively charged ions and free electrons, creating plasma with unique properties, including electrical conductivity and intense heat.

Plasma cutting leverages these properties to cut through metal. The equipment consists of a plasma torch, which is the heart of the operation. Within this torch, an electrical arc is generated between an electrode and the metal being cut. This arc elevates the temperature of a gas—commonly air, nitrogen, or argon—to such an extent that it transitions into a plasma state.

As the gas passes through a narrow opening in the torch, it is constricted into a focused stream of plasma. The electrical conductivity of the plasma allows the electrical arc to transfer to the workpiece, where the intense heat of the plasma (which can reach temperatures up to 30,000°C or higher) melts the metal. The force of the plasma jet then blows the molten metal away from the cut, creating a clean separation.

The process is facilitated by a power supply that provides a constant flow of DC electricity, maintaining the plasma arc throughout the cutting operation. The power supply's capacity to modulate current levels allows for adjustments in the intensity of the plasma, enabling cuts of varying depths and on different types of metal.

One of the unique aspects of plasma cutting is its ability to cut conductive metals with remarkable speed and precision. Unlike mechanical cutting methods, plasma cutting reduces the heat-affected zone (HAZ) around the cut, minimizing the thermal distortion of the workpiece. This precision makes it an invaluable tool in industries where metal fabrication requires both speed and accuracy, such as automotive manufacturing, construction, and metal artistry.

Plasma cutting also distinguishes itself through its versatility. With the right settings and consumables, plasma cutters can pierce and cut metals of different thicknesses and types, including steel, stainless steel, aluminum, brass, and copper. This adaptability extends to various cutting forms, from straight-line cuts to intricate patterns, all achievable with a single tool.

The efficiency of plasma cutting does not end with the cutting process itself. Advances in technology have integrated computer numerical control (CNC) systems with plasma cutting equipment, allowing for automated, precise cuts based on pre-programmed designs. This integration has expanded the capabilities of plasma cutting, pushing the boundaries of what can be achieved in metal fabrication.

Understanding the science behind plasma cutting opens up a world of possibilities for those involved in metalworking. It bridges the gap between theoretical knowledge and practical application, providing a foundation upon which one can build skills and explore creative and industrial applications. As technology advances, the principles of plasma cutting remain a cornerstone of innovation in the manipulation and fabrication of metal materials.

Components of a Plasma Cutter

Plasma cutting, a sophisticated process utilized in metalworking, leverages the unique properties of plasma to cut through metal with precision and efficiency. At the heart of this process are three critical components: the power supply, the Arc Starting Console (ASC), and the plasma torch. Each plays a pivotal role in transforming electrical energy into a concentrated plasma stream capable of slicing through metal with ease.

The power supply is the foundation upon which plasma cutting systems operate. It converts electrical power from a standard AC source into a continuous DC voltage, ranging between 200 to 400 volts. This conversion is essential because plasma cutting requires a stable and constant DC power supply to maintain the plasma arc. The magnitude of the DC voltage directly influences the intensity and temperature of the plasma arc, dictating the cutting speed and the thickness of metal that can be cut. Moreover, the power supply regulates the current flow based on the material being cut and the desired cutting speed, ensuring optimal performance and efficiency during the cutting process.

Adjacent to the power supply in its importance is the Arc Starting Console (ASC), a component dedicated to initiating the plasma arc. The ASC generates a high-frequency, high-voltage spark. This spark provides the initial energy needed to ionize the air between the electrode and the workpiece, creating a path for the plasma arc to follow. The creation of this path is crucial, as it allows the DC current from the power supply to transfer to the workpiece through this ionized air, effectively starting the cutting process. The ASC operates at the very beginning of the cutting process and may also engage if the plasma arc extinguishes during cutting, ensuring a continuous operation without manual re-ignition.

At the forefront of the cutting process is the plasma torch, a meticulously designed tool that directs the plasma arc onto the metal. The plasma torch consists of several key components, including the electrode, nozzle, swirl ring, and shielding cap, each contributing to the stability and focus of the plasma arc. The electrode, positioned inside the torch, conducts electricity from the power supply, while the nozzle focuses the plasma arc onto the metal. The swirl ring, aptly named, swirls the gas around the electrode, aiding in stabilizing the plasma arc and cooling the torch. The shielding cap protects the torch from molten metal and helps shape the plasma arc for optimal cutting performance.

The plasma torch's design is integral to its function, requiring precision engineering to withstand the extreme temperatures and conditions of plasma cutting. It utilizes compressed air or another gas, such as nitrogen or oxygen, depending on the material being cut. This gas is ionized into plasma inside the torch and then expelled at high velocity through the nozzle, creating the cutting action. The type of gas used can affect the quality of the cut, the cutting speed, and the consumable life, making the choice of gas an important consideration in the cutting process.

Together, these components form a system that harnesses the power of ionized gas to cut through metal with remarkable precision and speed. The power supply, Arc Starting Console, and plasma torch are not just parts of a machine but are the core elements that enable the plasma cutting process, each contributing to the efficiency, reliability, and quality of the cuts produced. Understanding the roles and functions of these components is essential for anyone looking to master the art and science of plasma cutting.

Types of Plasma Cutters

Plasma cutting technology has revolutionized the way industries and artisans alike work with metal, offering a versatile and efficient method for cutting through various types of material with precision and speed. At the core of this innovation are plasma cutters, which come in various forms to cater to a wide range of applications, skill levels, and complexity in cutting tasks. Two primary categories of plasma cutters dominate the market: manual plasma cutters and automated or Computer Numerical Control (CNC) plasma cutters. Each type serves distinct purposes, offering different levels of automation, precision, and capabilities to meet the needs of users ranging from hobbyists to large-scale industrial operations.

Manual Plasma Cutters

Manual plasma cutters are designed for hand-held operations, making them highly versatile and portable. They are the tool of choice for repair work, maintenance tasks, and any project where the cutter needs to be brought to the material. With a manual plasma cutter, the operator directly controls the cutting torch, guiding it along the metal to make the cut. This direct control allows for flexibility and adaptability in cutting various shapes and sizes but requires a steady hand and some skill to achieve clean, precise cuts.

The simplicity and portability of manual plasma cutters make them exceptionally popular among hobbyists, small workshops, and on-site repair jobs. They are available in various sizes and power ranges, accommodating everything from thin sheet metal to thicker plates. The key to mastering manual plasma cutting lies in developing a steady hand and learning to control the speed and angle of the cut, skills that can be enhanced with practice and experience.

Automated/Computer Numerical Control (CNC) Plasma Cutters

On the other end of the spectrum are automated or CNC plasma cutters, which represent the pinnacle of precision and efficiency in plasma cutting. These sophisticated machines are typically integrated into a table or a larger cutting system, where the plasma torch is mounted on a gantry and controlled by computer software. This setup allows for extremely precise cuts, with the ability to produce complex shapes, detailed designs, and repeat patterns with high accuracy and repeatability.

CNC plasma cutters are ideal for industrial applications, fabrication shops, and businesses where production speed, precision, and the ability to cut complex designs are paramount. The software component of CNC plasma cutters allows for the pre-programming of designs, significantly reducing the potential for human error and ensuring consistent results across multiple pieces. While the upfront cost and operational complexity of CNC plasma cutters are higher than their manual counterparts, the benefits in terms of productivity, precision, and the ability to automate the cutting process make them an invaluable asset for high-volume or precision-focused operations.

Understanding Plasma Cutters

The fundamental technology behind both manual and CNC plasma cutters is the same. They work by sending an electric arc through a gas that is passing through a constricted opening, which turns the gas into plasma. This plasma is then directed out of the nozzle to cut through metal by melting it at the point of contact. The choice between a manual and a CNC plasma cutter depends on the specific needs of the user, including the type of work being done, the level of precision required, the volume of work, and the budget.

Manual plasma cutters offer the advantages of portability and flexibility, ideal for on-site repairs, small projects, or hobbyist work. In contrast, CNC plasma cutters are suited for industrial applications and situations where precision and repeatability are critical. Regardless of the type, understanding the operation, capabilities, and safety requirements of plasma cutters is essential for achieving optimal results and maintaining a safe working environment.

Chapter: 2 Setting Up Your Plasma Cutter

Initial Setup and Safety Checks

Setting up a plasma cutter involves a series of critical steps and safety checks to ensure a smooth, safe operation from start to finish. This initial setup is not just about assembling parts and turning on switches; it's about creating a safe working environment that aligns with the high-energy nature of plasma cutting. Proper setup is foundational to achieving precise cuts while minimizing risks to the operator and surrounding area.

The process begins with selecting an appropriate workspace. Plasma cutting requires a clean, dry, and well-ventilated area to manage the fumes and particles that are generated. The space should also be free of flammable materials and liquids to reduce the risk of fire. Ensuring adequate ventilation either through natural means or with the assistance of extraction fans or systems is crucial to maintain air quality and operator health.

Once the workspace has been prepared, the next step involves assembling the plasma cutter according to the manufacturer's instructions. This includes securely attaching the torch to the power supply and ensuring that all connections are tight and correctly positioned. Special attention should be paid to the grounding of the equipment. A proper ground connection is critical for safe operation, as it prevents electrical shocks and contributes to the overall stability of the cutting process.

Air supply connection and regulation come next. Plasma cutters require a clean, dry air source or gas to operate. Connecting the air compressor and setting it to the correct pressure according to the manufacturer's specifications is essential for optimal performance. It's also important to check for any leaks in the air supply lines as these can affect cut quality and operational safety.

Electrical safety is paramount in the setup process. Before plugging in the equipment, inspect all electrical cords and connections for damage or wear. The power source should match the plasma cutter's requirements, and the use of a dedicated circuit is recommended to avoid overloading. Once everything is connected, the plasma cutter should remain off until all setup steps are completed.

Personal protective equipment (PPE) must be prepared and readily available before starting the plasma cutter. This includes, but is not limited to, a welding helmet with proper shade, safety glasses, flame-resistant gloves, a long-sleeved shirt, pants without cuffs, leather boots, and hearing protection. Preparing PPE beforehand ensures that the operator is protected from ultraviolet radiation, sparks, metal particles, and noise from the moment the cutter is activated.

The final step in the initial setup is a comprehensive safety check. This involves reviewing all connections, ensuring the work area is secure, checking that all safety gear is in place, and verifying that fire extinguishing equipment is within reach. It also includes planning the placement of the workpiece on a suitable cutting surface, usually a metal cutting table that allows for sparks and molten metal to fall away safely.

By meticulously following these steps, operators can establish a safe, efficient environment for plasma cutting. The initial setup and safety checks form the bedrock of responsible operation, paving the way for precision cutting while safeguarding the operator and workspace from the inherent risks of plasma cutting. This thorough preparation not only enhances the quality of work but also instills a culture of safety that is vital in any metalworking endeavor.

Assembling the Plasma Cutter

Assembling a plasma cutter is a crucial step that bridges the gap between acquiring this powerful tool and harnessing its potential to cut through metal with precision and efficiency. This process involves careful preparation, a thorough understanding of the components, and meticulous attention to detail to ensure that the equipment is set up safely and effectively.

The assembly begins with unboxing the plasma cutter and all its components. Typically, this includes the main unit, which houses the power supply and controls; the plasma torch, which is the handheld device that emits the plasma arc; grounding clamps; air hoses; and possibly consumables such as electrodes and nozzles, depending on the model.

The first step in the assembly process is to secure the main unit of the plasma cutter in a stable and accessible location. This location should be a clean, dry, and well-ventilated area to minimize the risk of contamination and overheating. It's important that the unit is placed close enough to the work area to allow for ease of movement, but far enough away to be protected from sparks and debris.

Next, the grounding clamp must be connected to the plasma cutter. This clamp is crucial for safety and performance, as it completes the electrical circuit and helps to stabilize the plasma arc. The grounding clamp should be attached to a clean, metal part of the workpiece or to the workbench, ensuring a solid, metal-to-metal connection free of paint, rust, or coatings that could impede electrical conductivity.

Following this, the air supply needs to be set up. Plasma cutting requires a steady supply of compressed air to create the plasma arc. This involves connecting the air compressor to the plasma cutter using the air hoses provided. It's important to check all connections for leaks and to ensure the air compressor is capable of delivering the correct pressure and flow rate required by the plasma cutter. The user manual will specify these requirements, and adjustments may be needed to align with these specifications.

The assembly of the plasma torch is a delicate process that requires attention to detail. This involves installing the consumables, including the electrode and nozzle, into the torch according to the manufacturer's instructions. These components must be seated correctly to ensure optimal performance and to prevent damage. Once the consumables are in place, the torch can be connected to the plasma cutter. This connection must be secure to allow for the control and power to flow to the torch.

Finally, before use, it's essential to perform a thorough check of all connections and settings. This includes verifying that the electrical connections are secure, the air supply is correctly attached and leak-free, and the grounding clamp is properly positioned. Additionally, setting the correct pressure on the air compressor and ensuring that the plasma cutter's settings (such as amperage) are adjusted for the specific cutting task at hand is crucial for achieving the best results.

Assembling a plasma cutter is not just about putting parts together; it's about setting the stage for safe and efficient operation. Each step, from positioning the main unit to connecting the air supply and assembling the torch, plays a critical role in the overall performance of the tool. By following these steps carefully and consulting the user manual, operators can ensure that their plasma cutter is assembled correctly, leading to smoother operation, cleaner cuts, and a safer working environment.

Connecting to a Power Source

Connecting a plasma cutter to a power source is a pivotal step in setting up your equipment for operation, requiring attention to detail to ensure both safety and optimal performance. Plasma cutters, sophisticated tools that use electric arcs to cut through metal with high precision, demand specific electrical conditions to function correctly. The process of connecting your plasma cutter to a power source encompasses understanding the electrical requirements of your cutter, preparing the electrical connection, and ensuring safety measures are in place.

Firstly, it's crucial to understand the electrical requirements of your plasma cutter, which are typically specified by the manufacturer. These requirements include the voltage and amperage needs of the device. Plasma cutters generally come in two main types based on their voltage requirements: those that operate on standard household 110-120V outlets and those that require a 220-240V power supply, often found in professional settings. The type of power supply your plasma cutter needs will dictate the kind of outlet you must use or install.

Before making any connections, inspect the power cord and plug of the plasma cutter for any damage. A damaged cord can be a significant safety hazard, leading to electric shocks or fire. If the cord or plug shows any signs of wear or damage, it must be replaced or repaired before use.

Once you have verified that the plasma cutter and its cord are in good condition, and you have identified the correct type of outlet, you can proceed to connect the plasma cutter to the power source. If you're using a cutter that requires a 220-240V outlet and you don't have one available, you may need to have one installed by a licensed electrician. This is not a step to take lightly, as improper installation can lead to serious risks, including fire or electrocution.

For plasma cutters that operate on a 220-240V supply, ensuring that the circuit is correctly rated for the cutter's amperage is essential. This often means that the cutter should be connected to a circuit with a dedicated breaker to avoid overloading the system, which can lead to circuit tripping and potential safety hazards. The specifics of this will vary based on the cutter's requirements and the existing electrical system, so consulting with an electrician is advised.

In addition to the correct voltage and amperage, grounding the plasma cutter is a critical safety step. Proper grounding helps to prevent electric shock by providing a path for electrical current to be safely dissipated into the earth in case of a malfunction. Ensure that the plasma cutter is connected to a grounded outlet and that the grounding mechanism of the cutter itself is functioning and not damaged.

Safety should always be your priority when connecting your plasma cutter to a power source. Always follow the manufacturer's instructions carefully. Wear rubber-soled shoes and ensure your work area is dry to reduce the risk of electric shock. Never attempt to modify the plug or use adapters to fit into an unsuitable outlet, as this can create a dangerous situation.

After ensuring that your plasma cutter is properly and safely connected to a power source, it's vital to test the setup under controlled conditions before beginning any cutting work. This involves turning on the cutter and observing its initial behavior without engaging the cutting process. Look for any unusual noises, smells, or behaviors that might indicate a problem. If everything seems to be operating as expected, you can proceed with the confidence that your plasma cutter is correctly set up and ready for use.

Connecting your plasma cutter to a power source is more than just plugging in a device; it's about understanding the electrical demands of your equipment, preparing your workspace accordingly, and prioritizing safety above all. With careful attention to these details, you can ensure a successful setup that allows you to harness the full potential of your plasma cutter, creating precise and clean cuts in metal projects.

Air Supply Connection and Regulation

The process of setting up a plasma cutter requires meticulous attention to various components, among which the air supply connection and its regulation stand out as pivotal. The plasma cutting process hinges on the use of compressed air or other gases to create the plasma arc necessary for cutting metal. This makes the proper connection and regulation of the air supply not just a matter of efficiency, but of safety and the quality of the cut as well.

To initiate the setup, one must first ensure the availability of a clean, dry air source. The importance of moisture control cannot be overstated; moisture in the air supply can significantly degrade the quality of the cut, reduce the lifespan of consumable components within the torch, and even damage the plasma cutter's internal systems. Therefore, integrating an air dryer or a moisture separator into the air supply line is a critical step. These devices work to remove moisture from the air, ensuring that the gas entering the plasma cutter is dry and clean.

Following moisture control, the focus shifts to connecting the air supply to the plasma cutter. This is typically done using high-quality, reinforced hoses that can withstand high pressures. It is crucial to check that all connections are secure and leak-free to maintain consistent air pressure and flow rate during operation. Leaks not only compromise the cutting process but can also pose safety risks.

Once the air supply is connected, the regulation of air pressure becomes the next critical step. Plasma cutters operate within a specific range of air pressures, which is usually detailed in the manufacturer's specifications. Operating with too low or too high air pressure can lead to poor cut quality and increased consumption of consumables. To manage this, a regulator is used. This device allows the operator to set the air pressure to the optimal level for the material being cut and the specific cutting requirements. The regulator should be adjusted while the plasma cutter is in operation, as the dynamic pressure can differ from static readings.

Furthermore, some plasma cutters feature built-in air pressure regulators and gauges that allow for easy monitoring and adjustment directly from the unit. For those that do not, an external regulator and gauge need to be installed on the air supply line. Ensuring the regulator is properly calibrated and functioning is essential for consistent and precise cuts.

In addition to pressure regulation, flow rate adjustment might also be necessary, especially when working with different materials or thicknesses. The flow rate can affect the temperature and velocity of the plasma arc, impacting the cut quality and speed.

Adjusting the flow rate, when possible, to suit the cutting task can lead to more efficient operations and better outcomes.

The role of the operator in regularly checking and maintaining the air supply system cannot be overlooked. This includes inspecting hoses for wear and damage, ensuring connections remain tight and leak-free, and regularly draining any moisture that collects in filters or separators. Such maintenance is crucial for sustaining the performance of the plasma cutter and extending its lifespan.

In summary, the air supply connection and regulation are foundational to the successful operation of a plasma cutter. Through diligent setup, regular maintenance, and precise adjustment of the air pressure and flow rate, operators can achieve optimal cutting results, maintain safety, and ensure the longevity of their equipment. This meticulous approach to air supply management underscores the technical proficiency required to harness the full potential of plasma cutting technology.

Chapter:3 Preparing to Cut

Material Preparation and Placement

In the domain of plasma cutting, preparing the material and its placement is as critical to the success of a cut as the operation of the plasma cutter itself. This phase lays the groundwork for the entire cutting process, ensuring that the final outcome meets the desired specifications for quality, precision, and safety.

The initial step in material preparation involves selecting the appropriate metal to be cut. Plasma cutters are versatile tools capable of slicing through a wide range of conductive metals including steel, stainless steel, aluminum, brass, and copper. However, the thickness, type, and condition of the metal can significantly affect the cutting process. It's essential to verify that the chosen metal is suitable for plasma cutting and that its thickness aligns with the capabilities of the plasma cutter at hand.

Once the material is selected, it's imperative to clean the surface thoroughly. Plasma cutting requires a clean, conductive path between the plasma cutter and the metal to ensure an efficient and stable cutting arc. Rust, paint, coatings, and any form of surface contaminants can impede this conductivity and interfere with the quality of the cut. Cleaning the metal surface with a wire brush or a suitable solvent can remove these impediments, enhancing the cutting process's effectiveness and the longevity of the cutter's consumables.

The next step is to mark the metal for cutting. Precision in plasma cutting not only depends on the steadiness and expertise of the operator but also on accurate markings that guide the cut. Using a chalk line, marker, or scribe, outline the desired cut on the metal surface. For intricate designs or patterns, templates or stencils can be invaluable. These aids ensure that the cuts are made exactly where needed, reducing the likelihood of errors and the waste of material.

Material placement is equally paramount in the preparation phase. The workpiece should be positioned securely on a stable, heat-resistant surface that allows for the metal to be fully supported while also permitting the plasma cutter's torch to move smoothly along the cut lines. Proper support for the metal not only prevents it from moving during the cutting process but also minimizes the risk of warping, bending, or creating incomplete cuts.

In addition, it's important to consider the environment around the cutting area. Ensure that there is ample space for the operator to move and manipulate the torch freely without any obstructions. The placement should also account for the direction of sparks and molten metal to avoid damage to surrounding areas or equipment. Utilizing a cutting table designed for plasma cutting can provide an ideal surface that supports the metal while allowing slag and sparks to fall away from the workpiece.

Safety should never be an afterthought during material preparation and placement. The cutting area should be free of flammable materials and equipped with appropriate ventilation or fume extraction systems to handle the smoke and fumes generated during the cutting process. Personal protective equipment, including gloves, eye protection, and hearing protection, should be worn at all times to safeguard against the hazards of plasma cutting.

By meticulously preparing the material and ensuring its proper placement, the foundation is set for a successful plasma cutting operation. This attention to detail not only maximizes the efficiency and accuracy of the cut but also enhances the safety and cleanliness of the work environment, leading to superior outcomes in any plasma cutting project.

Setting the Correct Amperage and Air Pressure

Setting the correct amperage and air pressure is pivotal in the operation of a plasma cutter, as these factors significantly influence the quality of the cut, the efficiency of the operation, and the lifespan of the consumables. The precision in adjusting both parameters stems from a deep understanding of how they affect the plasma cutting process and the ability to adapt to the material being cut and the desired outcome.

Amperage, or the current setting, directly correlates to the thickness of the metal being cut. Higher amperage settings are used for thicker materials to ensure the plasma arc can fully penetrate and cut through the workpiece. Conversely, lower amperage settings are suitable for thinner materials, preventing excessive heat that could warp or damage the metal. The key to setting the correct amperage lies in consulting the cutter's manual for the manufacturer's recommendations, which typically include a chart or guide specifying the appropriate amperage for various thicknesses and types of materials. It's important to start with these recommendations and adjust based on the observed performance. For example, if the cut is too slow or the quality is poor, increasing the amperage slightly may improve the results. However, exceeding the recommended amperage not only risks damaging the material but also wears out the consumables more rapidly.

Air pressure, measured in pounds per square inch (PSI), is another critical setting that affects the plasma cutting process. The compressed air not only helps to generate the plasma arc but also expels the molten metal from the cut. Setting the air pressure too low may result in an unstable arc and inadequate expulsion of molten material, leading to a poor quality cut. On the other hand, excessively high air pressure can lead to unnecessary wear on the consumables and might not improve the cut quality. Like amperage, the ideal air pressure setting depends on the thickness of the material and the specific plasma cutter being used. Most manufacturers provide recommended air pressure settings in the operator's manual, which should be considered a starting point. Observing the cut quality and adjusting the air pressure accordingly can fine-tune the process. A clean, narrow cut with minimal dross (residue) indicates that the air pressure is set correctly.

In practice, setting the correct amperage and air pressure requires a balance between following the manufacturer's guidelines and adjusting based on practical experience with the specific cutter and material. It's beneficial to conduct test cuts on scrap material of the same type and thickness as the workpiece to dial in the settings before proceeding with the final cut. This approach not only helps to conserve materials and consumables but also ensures the highest quality outcome for the project at hand.

Moreover, the interplay between amperage and air pressure settings underscores the importance of a holistic understanding of the plasma cutting process. Adjustments to one parameter may necessitate changes to the other, emphasizing the need for a methodical and attentive approach to preparation. By mastering these settings, operators can achieve optimal cut quality, efficiency, and consumable lifespan, thereby maximizing the capabilities of their plasma cutter.

Positioning and Securing the Workpiece

The process of positioning and securing the workpiece is a critical step in plasma cutting, ensuring not only the quality and precision of the cut but also the safety of the operator. Before the plasma cutter can be engaged, the metal being cut must be properly placed and stabilized to prevent movement that could result in inaccuracies or accidents. This preparation stage is where the foundation for a successful cut is laid, blending careful planning with practical safety measures.

The first step in positioning the workpiece involves selecting an appropriate surface or cutting table. This surface must be stable and capable of supporting the weight of the metal without risk of tipping or collapsing. Equally important, it should be constructed from a material that does not interfere with the plasma cutting process; for many, a grid or slat-based steel table is preferred, as it allows for the easy escape of cutting debris and heat.

Once a suitable surface is chosen, the next consideration is the placement of the workpiece itself. It should be laid out flat, with the area to be cut accessible and free of obstructions. If the design requires, it may be necessary to elevate the workpiece using metal stands or blocks, particularly if cuts need to be made from the underside or if the design involves intricate patterns that could be compromised by contact with the table surface. Care must be taken to ensure these supports are placed strategically to provide stability without becoming hazards themselves.

Securing the workpiece is paramount, as any movement during cutting can lead to imperfect cuts, damage to the metal or the cutting table, and increased risk of injury. Various clamps and fixtures are available for this purpose, designed to hold the metal firmly in place without obstructing the cutting path. Magnetic and C-clamps are commonly used due to their versatility and strength, but the choice of clamping device will depend on the specific requirements of the workpiece, including its size, shape, and thickness.

In addition to mechanical clamping, operators may also employ weights or other gravity-based methods to secure smaller or more delicate pieces. However, whatever method is used, it is crucial to verify that there is no undue stress on the metal that could cause warping or distortion during cutting. The aim is to maintain the natural state of the workpiece as much as possible to ensure accuracy.

The orientation of the workpiece relative to the plasma cutter should also be considered. The direction of the cut, the angle of the torch, and the position of the operator all play roles in determining the final quality of the cut. Planning these aspects in advance can minimize the need for adjustments during cutting, which, in turn, reduces the risk of errors or accidents.

Finally, it's essential to double-check the entire setup before beginning the cutting process. This includes a last review of the workpiece's security, the plasma cutter's settings, and the surrounding area to ensure it is free of unnecessary tools, debris, or other objects that could interfere with the operation.

By thoroughly addressing the positioning and securing of the workpiece, operators can create an optimal environment for plasma cutting. This preparation not only enhances the precision and efficiency of the cut but also reinforces safety measures, setting the stage for a successful and satisfying cutting experience.

Marking and Template Usage

Marking and template usage are critical steps in the preparation phase of plasma cutting, serving as the blueprint that guides the cutting process to achieve precise and accurate results. This preparatory stage ensures that the final cut reflects the intended design or measurement with minimal errors, optimizing both time and material efficiency.

Marking the workpiece involves a careful process of transferring your design or measurements onto the metal surface that is to be cut. This process begins with selecting the right marking tool. For most applications, soapstone, metal scribes, or special marking pens that are visible on metal and resistant to heat are preferred. These tools allow for clear, precise lines that can withstand the heat of plasma cutting without burning away or becoming obscured.

The accuracy of the marking process is paramount. It requires not only a steady hand but also an understanding of the cutting path. When marking, it's essential to consider the kerf, which is the width of the cut made by the plasma cutter. Adjusting the design to account for the kerf ensures that the final dimensions of the piece are accurate. For intricate designs or patterns, a template or stencil made from metal, plastic, or cardboard can be invaluable. Templates can be traced directly onto the workpiece, offering a repeatable pattern that ensures consistency, especially for multiple cuts of the same shape.

Template usage significantly streamlines the cutting process, particularly for complex designs or when mass-producing parts. Templates can be created using various methods, from manual drawing and cutting to sophisticated computer-aided design (CAD) software and CNC (Computer Numerical Control) machines that produce precise and intricate patterns. Regardless of the method used to create them, templates serve as a tangible guide that can be easily followed with the plasma cutter, reducing the margin for error and enhancing the quality of the cut.

In addition to traditional templates, modern plasma cutting systems may utilize digital or electronic guides. These systems can project a laser line onto the workpiece, outlining where the cut will be made. This method is particularly useful for complex or large-scale projects, providing a visual guide that can be adjusted before the actual cutting begins, thereby reducing waste and ensuring the accuracy of the final product.

Whether using manual marking tools or sophisticated digital guides, the goal remains the same: to transfer the intended design onto the metal in a way that is clear, precise, and adaptable to the cutting process. This preparation not only serves as a roadmap for the plasma cutter but also enables the operator to anticipate and adjust for potential issues, such as distortion from heat or the need for multiple passes on thicker materials.

In conclusion, the processes of marking and template usage are foundational to successful plasma cutting. They embody the meticulous preparation that precedes the power and precision of the plasma cutter, marrying the initial vision with the final creation. Through careful marking and the strategic use of templates, operators can achieve not just cuts that are precise and accurate, but also works of metal that are true to the envisioned design, whether in the realm of industrial fabrication, artistic creation, or custom manufacturing.

Chapter: 4 Operating the Plasma Cutter

Safety Precautions Before Cutting

Operating a plasma cutter, a tool renowned for its ability to slice through metal with precision and efficiency, necessitates a thorough understanding and adherence to safety precautions before the cutting process begins. These precautions are pivotal not only for the operator's safety but also for ensuring a smooth, efficient cutting experience.

Before initiating any cutting operation, it is imperative to conduct a comprehensive inspection of the plasma cutter and its components. This includes checking the torch, lead, and all connections for signs of wear, damage, or loose parts. Ensuring that the equipment is in optimal condition can prevent accidents caused by equipment failure. Additionally, verifying that the work area is clean, dry, and free of flammable materials minimizes the risk of fire and accidents, providing a safe environment for cutting.

Equally important is ensuring that the workpiece is securely fastened. A stable workpiece not only prevents accidents but also contributes to the precision of the cut. The grounding of the plasma cutter is a crucial step that cannot be overlooked. A properly grounded cutter reduces the risk of electric shock, a serious hazard given the electrical nature of plasma cutting.

Personal protective equipment (PPE) plays a critical role in the operator's safety. Prior to cutting, the operator must don all necessary PPE, which includes, but is not limited to, an auto-darkening welding helmet to protect against bright light and UV radiation, safety glasses or goggles to shield the eyes from debris, flame-resistant gloves to protect the hands from heat and cuts, and flame-resistant clothing to cover exposed skin. Additionally, ear protection guards against the potentially damaging noise levels, and respiratory protection is essential to avoid inhaling harmful fumes and dust.

Understanding the material being cut is also a fundamental safety precaution. Different materials can produce varying types of fumes and require specific settings on the plasma cutter to be cut safely and effectively. For instance, cutting materials like painted or coated metals may release toxic fumes, thereby necessitating adequate ventilation or even the use of a respirator.

Preparing the cutting area involves ensuring proper ventilation. The process of plasma cutting generates fumes and gases that can be harmful if inhaled. A well-ventilated area or the use of a fume extraction system is vital to maintain a safe breathing environment. Additionally, setting up fire extinguishers or having sand buckets nearby provides preparedness for any accidental fires.

Electrical safety is paramount. The operator should ensure that the plasma cutter and all auxiliary equipment are properly connected to a power source with the correct voltage and that all cables and cords are in good condition, without frays or cuts. Standing on a dry, insulated mat and wearing insulated boots can further mitigate the risk of electrical shock.

Lastly, before commencing the cutting operation, a final check of the immediate surroundings is essential to ensure that no one else is at risk from the cutting activity. This includes verifying that bystanders are at a safe distance and that they too are equipped with appropriate safety gear if they are to remain in the vicinity.

By meticulously following these safety precautions before cutting, operators can minimize the risks associated with plasma cutting, ensuring a safe work environment while harnessing the powerful capabilities of the plasma cutter to achieve precise and efficient cutting results. This diligent preparation underscores the respect and caution necessary when working with such a potent tool, reinforcing the principle that safety is paramount in all metalworking endeavors.

Igniting the Plasma Torch

Igniting the plasma torch is a pivotal step in the operation of a plasma cutter, marking the transition from preparation to the actual process of cutting metal. The method involves a carefully coordinated sequence of actions designed to safely and efficiently create a plasma arc capable of slicing through metal with precision and speed. Understanding this procedure in depth is crucial for anyone looking to master the use of a plasma cutter, as it lays the foundation for effective and safe cutting operations.

The ignition process begins with ensuring that the plasma cutter is correctly set up, connected to a suitable power source, and that the air supply is properly regulated. Once the equipment is verified to be in good working condition, the focus shifts to the plasma torch itself. The torch typically consists of several key components, including an electrode, a nozzle, and sometimes a swirling ring or gas distributor, which together facilitate the creation of the plasma arc.

To initiate the plasma arc, the operator must first press the trigger on the torch. In most modern plasma cutters, this action activates a pilot arc between the electrode and the nozzle inside the torch. This pilot arc is crucial as it ionizes the air passing through the torch, transforming it into a conductor of electricity. The presence of this ionized gas allows for the creation of the plasma arc when the torch is brought close to the workpiece.

The process of igniting the plasma arc involves a precise control over the distance between the torch tip and the metal surface. This distance, often referred to as the standoff distance, is critical for the maintenance of the arc. Too close, and the nozzle may touch the workpiece, risking damage to both the nozzle and the metal; too far, and the arc may fail to initiate or may extinguish shortly after being established. The optimal standoff distance varies based on the thickness of the metal and the specific model of the plasma cutter but generally ranges from a few millimeters to a centimeter.

Once the trigger is engaged and the pilot arc is established, the operator must smoothly move the torch towards the metal until the main plasma arc forms. This transition is marked by a noticeable increase in light and noise as the electrically charged gas cuts through the metal. It's essential for operators to maintain a steady hand and a consistent speed to ensure a clean cut, adjusting the movement based on the thickness of the material and the desired cut quality.

Safety during this phase cannot be overstated. Proper protective gear, including a welding helmet, gloves, and protective clothing, should be worn to shield the operator from the intense light, heat, and potential spatter produced during cutting. Additionally, maintaining awareness of one's surroundings and ensuring that the work area is clear of flammable materials and properly ventilated to remove fumes and smoke are fundamental to a safe working environment.

Igniting the plasma torch and establishing the plasma arc represent the core of plasma cutting operations. Mastery of this process requires not only an understanding of the technical aspects involved but also a commitment to safety and precision. Through practice and adherence to best practices, operators can leverage the power of the plasma cutter to achieve exceptional results, transforming raw materials into finished products with efficiency and artistry.

Techniques for Straight Cuts

Mastering the technique of making straight cuts with a plasma cutter is a foundational skill that significantly enhances the quality and precision of your metalwork. Achieving a perfectly straight cut requires understanding the equipment, the material you're working with, and applying a set of practiced techniques to ensure clean, precise lines.

The process begins before the cutter is even powered on, with preparation playing a crucial role. Ensure the metal surface is clean and free from any coatings, rust, or debris that could interfere with the cut. Secure the workpiece to prevent movement during cutting, as even slight shifts can result in deviations from a straight line.

One of the most effective strategies for guiding a plasma cutter along a straight path is to use a guide. A straight edge or metal ruler, clamped securely to the workpiece, serves as a physical guide for the torch. This guide should be made of a material that can withstand the heat and sparks generated during the cutting process. Align the guide so that the torch, when placed against it, follows the desired line of the cut.

When preparing to make the cut, position the torch at the starting point, ensuring that the tip or appropriate standoff distance is maintained. The standoff distance or height of the torch above the workpiece is critical for achieving a quality cut. Some plasma cutters come with a built-in guide that helps maintain this distance, while manual methods may involve using a spacer or roller guide.

Initiating the cut involves a steady hand and a smooth, consistent motion. Activate the plasma arc and begin moving the torch along the guide, keeping the torch perpendicular to the workpiece. The speed at which you move the torch is pivotal; moving too slowly can result in a wider kerf and excessive dross (melted metal that cools and solidifies on the edges of the cut), while moving too quickly can lead to an incomplete cut or diminished cut quality. The ideal speed achieves a steady stream of sparks shooting out from the bottom of the workpiece, indicating that the cut is penetrating fully through the metal.

For longer cuts, or to maintain consistent quality over extended periods, consider using a mechanized system or plasma cutting table equipped with guide rails. These systems can automate the cutting process, ensuring straight lines even over large distances. However, for manual cuts, practicing the technique, maintaining a steady hand, and keeping an eye on the torch's speed and angle are key.

In addition to physical techniques, there are also preparatory steps that can enhance the quality of straight cuts. For instance, marking the cut line with a soapstone or metal marking pen can provide a visual guide to supplement the physical guide. Also, familiarizing oneself with the plasma cutter's settings—such as amperage and air pressure—for the specific type and thickness of metal being cut can make a significant difference in cut quality.

In summary, making straight cuts with a plasma cutter is a skill that combines preparation, the use of guides, proper technique, and practice. It involves cleaning and securing the workpiece, using a straight edge to guide the torch, maintaining the correct standoff distance, moving the torch at an appropriate speed, and adjusting the cutter's settings to match the material. With patience and practice, achieving clean, straight cuts becomes a straightforward task, enhancing the overall quality and appearance of your metalwork projects.

Techniques for Curved Cuts and Holes

Achieving curved cuts and precision holes with a plasma cutter requires a blend of skill, practice, and understanding of the machine's capabilities. As plasma cutting becomes an indispensable tool in metalworking for its efficiency and versatility, mastering these techniques opens up a new realm of possibilities for intricate designs and detailed work.

The process of making curved cuts with a plasma cutter involves a steady hand and a keen eye for detail. The key is to maintain a consistent speed and distance between the torch and the metal surface. Moving too fast can result in incomplete cuts, while moving too slowly may cause excessive dross (molten metal) to form on the back of the cut. For beginners, practicing on scrap metal to get a feel for the machine's movement and response can be invaluable. As you guide the torch along the desired path, it's crucial to anticipate the cutter's behavior at curves and corners, adjusting your speed and direction smoothly to maintain the integrity of the curve.

Creating circles or curves requires a pivot point around which the plasma torch can rotate. This can be as simple as a nail or a screw placed at the center point of the desired circle or curve. Some cutters choose to use a compass-style guide, which can be either purchased or homemade, attaching to the torch and allowing for adjustable, consistent radius cuts. This tool is especially useful for repeat cuts or when precision is paramount.

Cutting precise holes, especially smaller ones, is more challenging due to the plasma arc's tendency to create a wider kerf at the beginning of the cut. To achieve clean, round holes, starting the cut with a piercing operation away from the edge of the intended hole is advisable. This method involves angling the torch at about 45 degrees to the work surface, allowing the arc to pierce the metal away from the final cut area, and then tilting it upright and moving into the hole's perimeter. This technique minimizes the potential for the arc to widen the entry point excessively.

For larger holes, a template or guide can be extremely useful. Templates made from metal can withstand the heat of the plasma cutting process and provide a guide to follow, ensuring the hole's circumference is even. Additionally, CNC (Computer Numerical Control) plasma cutters offer the highest degree of precision for holes and curved cuts, with computer programming guiding the torch in complex patterns that would be difficult or impossible to achieve by hand.

Regardless of the method, the quality of the cut can be affected by factors such as the type and thickness of the metal, the power settings of the plasma cutter, and the type of consumables used in the torch. Adjusting these variables to suit the specific cutting task can significantly improve the outcome.

After completing the cut, post-processing may be necessary to achieve the desired finish. This can include grinding or filing the edges to remove any dross or burrs, ensuring the final product is smooth and ready for use in the project.

In summary, cutting curves and holes with a plasma cutter is an art that combines technical knowledge with practical skill. By understanding the principles behind the equipment and applying these techniques, users can create intricate designs and precise cuts that elevate their metalworking projects.

Managing Speed and Movement

Mastering the management of speed and movement is a pivotal skill in operating a plasma cutter effectively. This expertise not only ensures the quality of the cut but also impacts the overall efficiency of the cutting process and the longevity of the plasma cutter's consumables. A well-calibrated balance between the cutter's speed and the operator's hand movement is essential for achieving precision and minimizing waste in any plasma cutting operation.

The optimal cutting speed is contingent upon several factors, including the thickness and type of material being cut, the specific settings of the plasma cutter, and the desired quality of the cut edge. Moving too slowly may result in excessive dross (molten material that cools and adheres to the bottom edge of the cut), excessive wear on the consumables, and even potential damage to the workpiece due to overheating. Conversely, moving too quickly can lead to incomplete cuts, an uneven cut surface, and a reduction in cut quality.

To determine the ideal speed for a particular task, operators should refer to the guidelines provided by the plasma cutter manufacturer, which often include recommended speed settings for cutting various materials and thicknesses. However, practical experience and experimentation also play a significant role in honing this skill. A good starting point is to begin cutting at the recommended speed and then adjust based on the results. For instance, if the cut is clean with minimal dross on the bottom edge, the speed is likely appropriate. If the dross is excessive, reducing the speed might be necessary.

The quality of the cut can also serve as a visual guide to managing speed. A perfectly optimized cut will leave a small, easily removable amount of dross, exhibit minimal bevel, and produce a narrow kerf. Observing the angle and brightness of the plasma arc can also provide cues; an optimal arc should be directed straight down into the material, indicating that the cutting speed is neither too fast nor too slow.

In addition to speed, the steadiness and consistency of the operator's movement play a crucial role in the quality of the cut. This is particularly true when cutting complex shapes or when precise angles are required. Using guides, templates, or CNC (Computer Numerical Control) systems can significantly improve consistency and accuracy. For manual cutting, practicing to maintain a steady hand and using tools like straight edges or roller guides can help in achieving smoother, more consistent cuts.

Furthermore, the technique of starting and ending the cut deserves special attention. Initiating the cut with a slight lead-in angle allows for a cleaner entry, while ending with a slight overrun ensures that the cut is completed without leaving any uncut sections. This practice not only enhances the appearance of the cut but also reduces the need for secondary finishing operations.

Understanding the interplay between speed and movement when operating a plasma cutter is an art that combines technical knowledge with tactile experience. Mastery in this area leads to improved cut quality, reduced material waste, and extended life of the plasma cutter's consumables. It underscores the importance of a thoughtful and deliberate approach to plasma cutting, where precision in speed and movement is not just a matter of efficiency but an expression of craftsmanship.

Troubleshooting Common Issues

Operating a plasma cutter, while highly efficient for cutting through various types of metal, can sometimes present challenges that impede performance. Recognizing and troubleshooting these common issues are essential skills that ensure smooth operation and high-quality cuts. This comprehensive exploration covers several frequent problems, their causes, and the steps to resolve them, facilitating uninterrupted use of the plasma cutter.

One prevalent issue is the difficulty in initiating the plasma arc. This problem often stems from an insufficient power supply or incorrect setup. Ensuring that the plasma cutter is connected to an appropriate power source and checking that all connections are secure and correctly configured can often resolve this issue. Additionally, verifying that the consumables are not worn out and are properly installed is crucial, as worn or incorrectly installed consumables can prevent the arc from starting.

Another common challenge is an unstable or fluctuating arc during cutting. This can be due to several factors, including improper air pressure, damaged consumables, or an inadequate grounding of the workpiece. Adjusting the air pressure to the manufacturer's recommended settings can help stabilize the arc. Regularly inspecting and replacing consumables as needed ensures that they perform optimally. Moreover, confirming that the workpiece is securely grounded eliminates potential electrical interference with the arc stability.

Dross formation on the cut edges is an issue that affects the quality of the cut, leaving behind slag that requires post-processing removal. This problem can be attributed to cutting speed, torch height, or incorrect amperage settings. Increasing the cutting speed can help reduce dross formation on thinner materials, while decreasing the speed is sometimes necessary for thicker materials. Adjusting the torch height to the recommended distance from the workpiece ensures optimal cutting conditions. Similarly, setting the amperage according to the thickness and type of metal being cut can improve cut quality and reduce dross buildup.

Poor cut quality, characterized by uneven or jagged edges, can also result from incorrect cutting speed or worn consumables. If the cutting speed is too fast, it may not allow the plasma arc to fully penetrate the material, leading to uneven cuts. Conversely, too slow a speed can cause excessive melting of the metal, resulting in jagged edges. Regular inspection and replacement of consumables are essential to maintain a precise and efficient cutting process.

Excessive consumption of consumables is another issue that affects the operational cost and efficiency of plasma cutting. This excessive wear can be caused by several factors, including operating the cutter beyond its intended capacity, incorrect cutting parameters, or poor quality consumables. Using the plasma cutter within its recommended settings and ensuring the use of high-quality consumables can significantly extend their life. Additionally, adhering to the recommended cutting parameters for each type of metal and thickness can optimize consumable usage.

Troubleshooting these common issues in plasma cutting requires a systematic approach to identifying and addressing the root causes. By understanding the operational principles of the plasma cutter and adhering to recommended practices for setup, maintenance, and operation, users can minimize downtime and maximize the efficiency and quality of their cuts. This proactive approach to troubleshooting not only enhances the functionality of the plasma cutter but also contributes to the longevity of the equipment, making plasma cutting a reliable and valuable method in metal fabrication and artistic endeavors.

Chapter: 5 Advanced Techniques

Piercing Thick Materials

Piercing thick materials with a plasma cutter is an advanced technique that demands a deep understanding of the equipment's capabilities and the physical properties of the material being cut. This process is pivotal for professionals who deal with industrial-grade metals and require precise, high-quality cuts through materials that are often challenging to penetrate. To achieve optimal results and maintain safety, operators must consider several critical factors, including the setup of the plasma cutter, the technique used for piercing, and the management of potential hazards.

The initial step in piercing thick materials involves ensuring the plasma cutter is appropriately configured for the task. This includes selecting a plasma cutter with a sufficient amperage rating for the thickness of the material. Generally, a higher amperage machine offers the capacity to cut through thicker materials more efficiently. Moreover, equipping the plasma cutter with the correct consumables is crucial, as these components—such as the nozzle, electrode, and shield—must be capable of handling the increased power and heat involved in piercing thick materials. Using consumables designed for high-amperage cutting can significantly improve the quality of the cut and extend the life of the plasma cutter.

The technique for piercing thick materials also differs from cutting thinner metals. Rather than starting the cut at the edge of the material, the operator must begin the piercing process away from the edges, in the area where the hole is desired. This approach requires holding the torch at a slight angle to the workpiece surface to direct the initial blast of plasma and molten metal away from the operator. Once the piercing begins, the operator can gradually move the torch to a vertical position, allowing the plasma arc to penetrate through the material. It's essential to maintain a steady hand and control the torch's motion to prevent damage to the consumables and ensure a clean pierce.

One of the key challenges in piercing thick materials is managing the substantial amount of molten metal and slag that the process generates. This not only poses a risk to the operator's safety but can also affect the quality of the pierce. To mitigate these issues, operators can employ a piercing shield or standoff guide. These accessories help to protect the torch and consumables from the molten metal, ensuring a safer and cleaner piercing process. Additionally, setting the correct air pressure and amperage on the plasma cutter can help control the amount of molten metal produced.

Another advanced technique involves preheating the material in the area to be pierced. Preheating can reduce the thermal shock to the material, facilitating a smoother piercing process and minimizing the risk of cracking or warping, especially in thicker, more temperamental materials. This method requires careful control of the heat source to avoid overheating the metal, which could compromise its structural integrity.

Operators must also be mindful of the potential hazards associated with piercing thick materials, such as the intense UV and infrared radiation generated by the plasma arc. Proper safety gear, including a high-quality welding helmet, gloves, and protective clothing, is imperative to protect against burns, eye damage, and exposure to harmful radiation. Adequate ventilation is equally crucial to remove toxic fumes and smoke from the workspace.

Piercing thick materials with a plasma cutter represents a sophisticated application of plasma cutting technology, requiring a blend of technical knowledge, precise technique, and adherence to safety practices. When executed correctly, this advanced technique allows for efficient and precise penetration of heavy-duty materials, opening up new possibilities for fabrication and construction projects. As with all advanced plasma cutting techniques, continuous learning, practice, and respect for the equipment's power and potential hazards are key to mastering the art of piercing thick materials.

Beveling Edges

Beveling edges with a plasma cutter represents a sophisticated technique that elevates the skill set of any metalworker. This process involves cutting a sloped edge on a piece of metal, a crucial step in preparing metal for welding, creating precise joints, or adding aesthetic details to metal projects. Mastering beveling with a plasma cutter not only enhances the functionality and appearance of metalwork but also demonstrates a proficiency in handling one of the most versatile tools in the metal fabrication industry.

The essence of beveling lies in its ability to prepare metal pieces for a stronger, more seamless weld. By creating an angled surface, beveling allows for a larger surface area to be welded, resulting in a joint that is significantly stronger than one made with squared edges. Beyond practicality, beveled edges also contribute to the visual appeal of metal pieces, offering a refined finish that distinguishes professional work from amateur efforts.

To achieve precision in beveling with a plasma cutter, understanding the equipment's capabilities and settings is paramount. A plasma cutter equipped with a beveling head allows for the adjustment of the torch angle, enabling the operator to cut at the desired bevel angle. For machines without an adjustable head, manual positioning of the torch is required. This demands a steady hand and an eye for detail as the angle and speed at which the cutter is moved play critical roles in achieving a consistent bevel.

The thickness of the metal, the type of edge required (straight, curved, or irregular), and the specific angle of the bevel are key factors that influence the approach and settings for a beveling project. The plasma cutter's amperage settings need to be adjusted according to the metal's thickness, with higher amperage necessary for thicker materials to ensure a clean, smooth cut. The choice of consumables, particularly the nozzle and electrode, also impacts the quality of the bevel, requiring selections that match the cutter's specifications and the demands of the project.

Technique further refines the beveling process. For straight edges, using a guide that maintains the torch at a constant angle relative to the workpiece can improve accuracy. For curved edges, freehand cutting demands a well-practiced technique to maintain a consistent angle and speed. The initiation of the cut, the movement across the material, and the completion of the cut all require attention to detail to prevent irregularities in the bevel.

One of the advanced skills in beveling with a plasma cutter involves managing the kerf, the width of the cut made by the cutter. Understanding how the kerf width changes with different angles and speeds is essential for achieving precise bevel angles, particularly when fitting parts together for welding. Adjusting the torch speed to control the amount of heat applied to the metal can minimize the kerf and produce cleaner, sharper beveled edges.

Post-processing is often necessary to perfect the beveled edge. This may include grinding or sanding to remove any dross (slag) left by the cutting process and to smooth the bevel for welding or aesthetic purposes. Such finishing touches enhance the fit-up of welded pieces and contribute to the overall quality and appearance of the metalwork.

In summary, beveling with a plasma cutter is an art that combines technical knowledge with skilled execution. It requires an understanding of the equipment, material properties, and cutting techniques, as well as practice and precision. Mastering this advanced technique not only broadens the range of projects that can be undertaken but also elevates the quality and craftsmanship of the work produced, showcasing the capabilities of the modern metalworker.

Cutting Stacked Layers

Cutting stacked layers with a plasma cutter is an advanced technique that allows operators to efficiently cut through multiple sheets of metal simultaneously. This method not only saves time but also ensures uniformity and consistency in cuts across several pieces. However, to achieve optimal results and maintain safety, it's crucial to understand the intricacies involved in stacking and cutting layers of metal with a plasma cutter.

Firstly, selecting the right materials is essential. The metals being stacked should be of the same type and thickness to ensure an even cut. Variations in material type or thickness can lead to uneven cuts, increased wear on the plasma cutter's consumables, and potential safety hazards. Preparing the metal sheets by cleaning them and ensuring they are free from coatings, rust, or debris is crucial for a clean cut and to reduce the risk of fire or harmful fumes.

The stacking process requires precision. Sheets should be aligned perfectly to avoid shifts during the cutting process. Clamping the stack securely is vital to prevent movement, which can cause inaccurate cuts or damage to the plasma torch. For thicker stacks, more powerful plasma cutters with higher amperage settings are needed. It's important to consult the plasma cutter's manual to determine the maximum stack thickness it can handle effectively.

When setting up the plasma cutter, adjusting the amperage and air pressure according to the total thickness of the stack is crucial. Higher amperage settings are required for thicker stacks to ensure the plasma arc penetrates through all the layers. However, too high an amperage setting can lead to excessive dross (molten metal) on the bottom layer and increased wear on consumables. Finding the right balance is key to efficient and clean cuts.

The cutting technique also needs to be adjusted when cutting stacked layers. A slower cutting speed is often necessary to allow the plasma arc to fully penetrate through all the layers. Starting the cut on an edge rather than in the middle of the sheet can also help ensure complete penetration from the outset. Operators should monitor the cut closely to adjust the speed as needed, ensuring the arc maintains its effectiveness through all layers.

In terms of safety, cutting stacked layers presents additional hazards. The increased amount of molten metal and sparks requires operators to wear full protective gear, including a welding helmet, gloves, and apron. Proper ventilation or a fume extraction system is also more critical due to the larger volume of fumes produced.

After cutting, the layers should be carefully separated and inspected. The top layers may cool faster than the bottom ones, so it's important to handle the stack carefully to avoid burns.

Inspecting the cut quality on each layer is necessary to ensure the technique was effective across the entire stack.

Cutting stacked layers with a plasma cutter is a valuable technique for operators looking to increase productivity and ensure uniformity in their work. However, it requires a thorough understanding of the plasma cutting process, precise setup, and careful execution. With practice and attention to detail, operators can master this advanced technique, making it a powerful addition to their metalworking arsenal.

Utilizing CNC Plasma Cutting for Precision Work

Utilizing CNC (Computer Numerical Control) plasma cutting represents a significant leap forward in the realm of precision metal fabrication. This advanced technique marries the efficiency and power of plasma cutting with the meticulous accuracy and repeatability offered by computer-controlled operations. CNC plasma cutting systems transform raw materials into intricate designs, precise cuts, and detailed metal components with a level of precision and speed that manual plasma cutting can't match.

At the core of CNC plasma cutting is the plasma cutter, which is integrated into a computer-controlled system that directs the cutter's movement over the material based on a pre-programmed design. This system consists of a table that holds the workpiece securely in place, a plasma torch that moves along multiple axes, and a computer or controller that directs these movements based on digital designs. The software used in these systems allows for the creation or importation of complex designs, which are then translated into precise cutting paths by the CNC machine.

One of the primary advantages of CNC plasma cutting is its unmatched precision. The computer-controlled setup eliminates human error and ensures that each cut is exactly the same, which is crucial for mass production or in applications where uniformity and precision are paramount. Furthermore, the speed of the cutting process is significantly enhanced. Once the design is programmed into the controller, the CNC plasma cutter can perform cuts much faster than manual methods, reducing production time and increasing efficiency.

CNC plasma cutting also excels in its versatility and capability to handle complex cuts and intricate designs. From detailed artistic projects to precise industrial components, the system can manage a wide range of cutting tasks without the need for extensive manual intervention. This capability opens up new possibilities for designs that would be too time-consuming or difficult to execute manually.

Another significant aspect of CNC plasma cutting is its optimal material utilization. The precision of the system minimizes waste by maximizing the number of parts that can be cut from a single piece of material. This efficiency not only reduces material costs but also contributes to sustainable manufacturing practices by minimizing scrap.

Despite its many advantages, mastering CNC plasma cutting requires a deep understanding of both the software and the hardware involved. Operators must be proficient in CAD (Computer-Aided Design) software to create or modify designs and CAM (Computer-Aided Manufacturing) software to translate those designs into commands the CNC machine can execute. Understanding the properties of different metals and how they interact with the plasma cutting process is also crucial, as this knowledge allows for adjustments in cutting speed, amperage, and torch height to optimize cut quality and equipment life.

Maintenance plays a vital role in ensuring the longevity and performance of a CNC plasma cutting system. Regular checks and replacements of consumable parts in the plasma torch, along with periodic calibration of the system, help maintain cutting precision and prevent downtime.

In conclusion, CNC plasma cutting is a sophisticated technique that brings unparalleled accuracy, speed, and efficiency to metal cutting. Its ability to produce precise cuts and intricate designs across a wide range of materials makes it a valuable tool in industries ranging from aerospace to art. However, to fully leverage the capabilities of CNC plasma cutting, operators must have a solid foundation in both the technology and techniques involved, underscoring the importance of comprehensive training and experience in this advanced method of plasma cutting.

Chapter: 6 Maintenance and Troubleshooting

Routine Maintenance Checks

Routine maintenance checks are a pivotal aspect of ensuring the longevity, efficiency, and safety of a plasma cutter. Regular upkeep not only prevents unexpected downtimes but also optimizes performance, ensuring that each cut is precise and clean. Given the sophisticated technology behind plasma cutters, a systematic approach to maintenance is essential.

The first step in routine maintenance involves the inspection and cleaning of the plasma cutter's exterior. Dust, debris, and any signs of wear or damage should be addressed immediately. A clean machine is less likely to overheat and more likely to operate efficiently. The air intake vents should be clear of obstructions to facilitate proper cooling of the internal components.

The electrical connections warrant close attention next. All connections should be tight and free of corrosion. Loose or corroded connections can lead to poor performance or even pose a safety risk. It's also crucial to inspect the power cord and plug for any damage that might compromise the safe operation of the device.

Air flow and air quality are critical to the performance of a plasma cutter. The air filter should be inspected regularly and replaced as needed to ensure clean, dry air is supplied to the system. Moisture in the air supply can damage the internal components of the plasma cutter and degrade the quality of the cut. Some systems are equipped with a moisture separator, which should be drained regularly to remove any accumulated water.

The condition of the consumables—such as the electrode, nozzle, retaining cap, and swirl ring—is perhaps the most critical aspect of maintenance. These components wear out over time and can significantly impact the performance of the plasma cutter. Regular inspection and replacement of worn consumables are essential. A rule of thumb is to check these components daily if the plasma cutter is used frequently. Signs of wear include uneven or excessive electrode wear, nozzle enlargement, or damage to the swirl ring. Using worn consumables not only decreases the quality of the cut but can also increase the operating cost and risk damage to the torch.

The torch itself requires careful maintenance. Inspect the torch body and lead for signs of wear, damage, or contamination. The torch coolant level (in systems that use a liquid-cooled torch) should be checked and topped up as necessary to ensure optimal cooling and performance.

For plasma cutters equipped with a CNC (Computer Numerical Control) interface, software updates and calibrations may periodically be necessary. Keeping the software up to date can enhance functionality and prevent operational issues.

After each use, a quick wipe-down of the machine and a visual inspection can help catch potential problems early. Additionally, maintaining a maintenance log can help track the performance of the plasma cutter over time and identify recurring issues that may require a more in-depth investigation or professional servicing.

In conclusion, routine maintenance checks are an indispensable part of using a plasma cutter effectively. They extend the lifespan of the machine, ensure operational safety, and maintain the quality of the cuts. Adhering to a regular maintenance schedule and promptly addressing any issues that arise are key to getting the most out of a plasma cutting system.

Replacing Consumables

In the realm of plasma cutting, maintaining optimal performance of your equipment is crucial for ensuring the precision and quality of your cuts, as well as for upholding safety standards. One integral aspect of this maintenance involves the timely replacement of consumables. Consumables are the parts of the plasma cutter that bear the brunt of the cutting process and, therefore, wear out over time. These include the electrode, nozzle, retaining cap, swirl ring, and shield.

Understanding when and how to replace these consumables is essential for anyone looking to master the use of a plasma cutter. The lifespan of consumables can vary significantly based on several factors, including the cutting amperage, the material being cut, and the duration and frequency of cutting. Consistently monitoring these parts for signs of wear or damage is the first step in ensuring your plasma cutter operates effectively and safely.

The electrode and nozzle are among the first components to show wear. The electrode is responsible for creating the plasma arc, and over time, it will begin to form a pit from the constant exposure to intense heat. Once this pit reaches a certain depth, the electrode's performance diminishes, affecting the quality of the cut. Similarly, the nozzle can become worn or clogged, leading to a less focused arc and, consequently, less precise cuts. Visual inspection after each use can help identify these wear signs early.

The retaining cap, swirl ring, and shield also play critical roles in the plasma cutting process. The retaining cap holds the consumables in place, while the swirl ring ensures the plasma gas swirls around the electrode correctly, and the shield protects the torch from sparks and molten metal. Wear on these parts can lead to a range of issues, from poor cut quality to damage to the torch itself. Regular inspection will help identify when these parts need to be replaced.

Replacing consumables is a straightforward process, but it requires careful attention to detail. Always refer to the manufacturer's guidelines for your specific model to ensure compatibility and correct installation. Generally, the process involves turning off and unplugging the plasma cutter, removing the torch's outer casing, and then replacing the worn consumables with new ones. It's crucial to handle these parts carefully and to ensure they are fitted securely and correctly to avoid any issues during operation.

Proper care and handling of consumables extend their life and the overall efficiency of your plasma cutter. For example, regularly cleaning the torch and consumables with compressed air can prevent the buildup of dust and debris, which can cause premature wear. Additionally, using the correct amperage settings for the material being cut can prevent excessive wear on the consumables.

Keeping a stock of replacement consumables on hand is also wise, ensuring that you can quickly replace worn parts and minimize downtime. Investing in high-quality consumables can also lead to longer life spans and better performance, although they may come at a higher initial cost.

In summary, the diligent replacement and maintenance of consumables are paramount in the effective operation of a plasma cutter. By understanding the signs of wear, conducting regular inspections, and following the manufacturer's guidelines for replacement, users can ensure their plasma cutter performs at its best. This not only leads to higher quality cuts but also extends the life of the plasma cutter and enhances safety during its use.

Troubleshooting Cutting Problems

Troubleshooting cutting problems in plasma cutting is an essential skill that operators must develop to ensure smooth operation and maintain the quality of their work. Plasma cutters, while efficient and versatile, can sometimes present issues such as uneven cuts, excessive dross (slag), poor cut quality, or difficulties in piercing thick materials. Understanding the root causes of these problems and knowing how to address them can significantly reduce downtime and improve cutting performance.

One common issue encountered is an uneven or jagged cut edge, often resulting from incorrect cutting speed. If the speed is too fast, the plasma arc may not fully penetrate the material, leading to an incomplete cut. Conversely, cutting too slowly can cause excessive dross and a wider kerf (cut width). The key is finding the optimal speed for the material thickness and the specific plasma cutter being used. Regularly consulting the cutter's manual for recommended settings and experimenting with slight adjustments can help achieve a smoother cut.

Excessive dross or slag on the bottom edge of the cut is another frequent problem. This is typically due to one of several factors: cutting speed, as previously mentioned; too high or too low of an amperage setting; or incorrect standoff distance—the gap between the torch tip and the workpiece. Ensuring the correct parameters and maintaining a consistent standoff distance can markedly improve cut quality and reduce the need for post-cut cleanup.

Poor cut quality can also result from worn or damaged consumables within the torch itself. The consumables, including the electrode, nozzle, and shield, play critical roles in focusing and maintaining the plasma arc. Over time or with extensive use, these components can wear out, leading to diminished cut quality and increased operational inefficiencies. Regular inspection and timely replacement of consumables are crucial maintenance steps. Keeping a log of consumable life can help predict when changes are needed and prevent quality issues before they start.

Difficulty piercing thick materials without causing excessive dross or enlarging the pierce hole beyond acceptable tolerances can challenge even experienced operators. One strategy to mitigate this issue is to initiate the pierce slightly off the edge of the material, creating a lead-in cut. This technique can reduce the amount of molten metal blowback onto the torch and improve the overall appearance of the pierce point. Additionally, using a machine with a high-frequency starting feature can help establish the arc without contacting the material, reducing wear on consumables and improving piercing performance.

Another aspect of troubleshooting involves the plasma system's electrical components. Issues such as an erratic arc or difficulty maintaining an arc can be indicative of electrical problems. These might include loose connections, damaged leads, or problems within the power supply itself. Conducting regular inspections and maintenance checks, including verifying all connections are secure and inspecting cables for damage, can help identify and rectify these issues.

Lastly, the quality of the air supply is a critical yet often overlooked factor affecting plasma cutter performance. Moisture, oil, or particulate matter in the compressed air can lead to various cutting problems, including poor cut quality and premature wear of consumables. Installing an air filtration or drying system can significantly improve air quality, and regularly draining moisture from the compressor tank and air lines can mitigate these issues.

In summary, effective troubleshooting of plasma cutting problems involves a holistic understanding of the cutting process, including machine settings, consumable condition, material properties, and external factors like air quality. By systematically addressing each potential issue area, operators can enhance their plasma cutting operations, reduce downtime, and achieve consistently high-quality cuts.

Dealing with Electrical Issues

Dealing with electrical issues in plasma cutting systems necessitates a proactive approach, combining regular maintenance with adept troubleshooting to ensure both the longevity of the equipment and the safety of its users. Plasma cutters, sophisticated tools that they are, depend on a stable and efficient electrical system to generate the plasma arc needed for cutting through metal. When electrical problems arise, they can not only hinder the cutting process but also pose significant safety risks.

The foundation of preventing electrical issues lies in understanding the electrical components of a plasma cutter. These include the power supply, which converts electrical energy into a suitable form for creating the plasma arc; the arc starting console (ASC), which initiates the arc; and the torch itself, which houses the electrode and nozzle critical for directing the plasma arc. Each component must function correctly and in harmony with the others to ensure efficient and safe operation.

Regular maintenance is the first line of defense against electrical problems. This includes inspecting and cleaning the torch components, checking all electrical connections for tightness and signs of wear, and ensuring the power supply is free of dust and debris that could impede its operation. Manufacturers often provide specific maintenance schedules and procedures, which should be followed diligently.

When electrical issues do arise, effective troubleshooting becomes crucial. One common problem is difficulty in starting the arc, which could be due to several factors, including a worn-out electrode, incorrect assembly of the torch parts, or issues within the ASC. Replacing the electrode, ensuring the torch is correctly assembled according to the manufacturer's guidelines, and checking the ASC for any signs of malfunction can often resolve these issues.

Another frequent issue is an unstable or fluctuating arc. This can be caused by insufficient power supply, incorrect air pressure, or problems with the torch components, such as a worn nozzle. Verifying that the power supply meets the cutter's requirements, adjusting the air pressure to the recommended setting, and replacing any worn torch parts can help stabilize the arc.

Electrical shocks, while less common, are a serious concern that can be caused by improper grounding, damaged insulation, or moisture in the work environment. Ensuring the plasma cutter and workpiece are properly grounded, regularly inspecting cables and connectors for damage, and keeping the work area dry are critical steps in preventing electrical shocks.

Overheating is another issue that can arise, particularly during extended use of the plasma cutter. This may indicate an overburdened power supply or insufficient cooling. Ensuring the plasma cutter's cooling system is functioning properly and avoiding exceeding the duty cycle of the equipment can prevent overheating.

In dealing with electrical issues, the importance of referring to the plasma cutter's manual cannot be overstated. The manual typically includes a troubleshooting section that provides guidance specific to the model, including how to diagnose and rectify common electrical problems. When in doubt, or if an issue persists, contacting the manufacturer's technical support team or a qualified technician is advisable.

By adhering to a regimen of regular maintenance, being vigilant for signs of electrical issues, and applying targeted troubleshooting strategies, users can ensure their plasma cutting equipment remains in optimal working condition. This proactive approach not only extends the lifespan of the plasma cutter but also upholds the highest standards of safety and efficiency in metal cutting operations.

Chapter: 7 Applications of Plasma Cutting

Industrial Applications

Plasma cutting, with its unmatched speed and precision, has revolutionized the way industries cut metal, becoming an indispensable tool across various sectors. This technology harnesses the power of an electrically ionized gas—plasma—to melt and precisely cut through metal, offering a level of efficiency and accuracy that traditional cutting methods struggle to match. Its versatility and efficiency have made it a cornerstone in the industrial landscape, finding applications in fields as diverse as manufacturing, automotive repair, construction, and even art.

In the manufacturing sector, plasma cutting is pivotal in the production of machinery and equipment. Its ability to swiftly cut through thick sheets of metal with minimal waste makes it ideal for creating large components required in machinery. Manufacturers value plasma cutting for its speed and precision, which significantly reduce production times and costs. Furthermore, the technology's adaptability allows for cutting various metal types and thicknesses, making it suitable for producing a wide range of industrial machinery and equipment.

The automotive industry benefits from plasma cutting in both manufacturing and repair. In manufacturing, it's used to cut intricate shapes and parts with high precision, essential for the complex designs of modern vehicles. For repair work, plasma cutters are prized for their ability to make quick and precise cuts in damaged metal parts, facilitating efficient repairs and restorations. This is particularly valuable in custom automotive shops, where precision and quality are paramount.

Construction is another field where plasma cutting has made significant inroads. Its ability to cut through structural steel has revolutionized the way buildings and infrastructures are constructed. Plasma cutters are used to shape steel beams, cut pipes, and prepare metal sheets, enabling complex architectural designs and sturdy constructions. Its precision and speed also mean that construction projects can progress faster, with materials prepared exactly to specification, reducing both time and labor costs.

The versatility of plasma cutting extends beyond functional applications to the realm of art and design. Artists and designers use plasma cutters to create intricate metal artworks, sculptures, and decorative pieces. The tool allows for precise cuts and creative freedom, enabling artists to work with metal in ways that were not possible with traditional cutting methods. This has given rise to a new genre of metal art, characterized by detailed and intricate designs that showcase the precision of plasma cutting.

Shipbuilding and aerospace industries also leverage plasma cutting technology. In shipbuilding, it is used to cut through thick plates of steel, essential for constructing the hulls of ships and submarines. The aerospace industry, known for its stringent requirements for precision and quality, utilizes plasma cutting to fabricate components from high-strength alloys. These components must withstand extreme conditions, and plasma cutting ensures they are cut to exact specifications.

In addition to these, plasma cutting finds applications in the fabrication of agricultural machinery, the production of heating, ventilation, and air conditioning (HVAC) components, and the metal furniture industry. Its ability to deliver clean cuts and detailed work without warping the material makes it an invaluable tool across these sectors.

The industrial applications of plasma cutting are vast and varied, underpinned by the technology's key attributes of speed, precision, and versatility. As industries continue to evolve and demand more efficient and precise manufacturing methods, the role of plasma cutting is set to grow, further cementing its status as a critical technology in the modern industrial landscape. Its impact spans from the functional to the aesthetic, demonstrating that plasma cutting is not just a method of cutting metal but a tool that shapes the very infrastructure of our world and the art that enriches it.

Artistic and Decorative Uses

Plasma cutting, a process known for its precision and efficiency in cutting through metal, has transcended its industrial origins to become a favored tool among artists and designers. This transformative use of plasma cutting technology has opened up a new realm of possibilities for artistic and decorative applications, enabling creators to work with metal in ways that were once difficult or impossible. By harnessing the power of a plasma cutter, artists and craftspeople can bring intricate designs to life, from stunning sculptures to delicate jewelry, and everything in between.

One of the most compelling aspects of using a plasma cutter for artistic purposes is its ability to cut with incredible detail. Unlike traditional metalworking tools that may struggle with fine or complex cuts, a plasma cutter can easily navigate tight curves and sharp angles, making it ideal for creating intricate patterns and designs. This precision allows artists to work on a granular level, etching detailed motifs into metal surfaces or cutting out delicate shapes for use in larger compositions.

Sculpture is a field where plasma cutting has had a profound impact. Artists can cut thick metal plates with ease, layering them to create depth and texture or bending them to form flowing, organic shapes. These techniques enable the creation of large-scale sculptures that are both durable and dynamic, capable of withstanding the elements when displayed outdoors while capturing the fluidity and movement of natural forms.

In the realm of decorative arts, plasma cutting has found its place in creating bespoke furniture, lighting fixtures, and architectural elements. Designers can cut elaborate patterns into panels of metal to be used in gates, railings, and screens, combining functionality with aesthetic appeal. These creations add a touch of elegance and uniqueness to homes, gardens, and commercial spaces, showcasing the versatility of metal as a decorative material.

Jewelry making is another area where the precise cuts achievable with a plasma cutter are invaluable. Artists can produce intricate pendants, earrings, and other adornments that stand out for their craftsmanship and originality. The ability to work with a variety of metals, from steel to copper, allows for a wide range of styles and finishes, catering to diverse tastes and preferences.

Signage and lettering are yet more applications where plasma cutting shines. Sign makers can produce custom signs with complex fonts and logos, cutting through metal with clean, crisp edges that paint or other finishing touches can further enhance. This capability opens up opportunities for businesses to brand their spaces uniquely and memorably.

Moreover, the democratization of plasma cutting technology, with more affordable and user-friendly models becoming available, has made these artistic endeavors more accessible to hobbyists and professionals alike. Workshops and maker spaces often offer access to plasma cutters, along with classes to teach the necessary skills, fostering a community of creators who share techniques and inspiration.

In conclusion, the artistic and decorative applications of plasma cutting are as varied as they are impressive. By offering the ability to work with metal in new and innovative ways, plasma cutting has become a staple in the toolkit of artists, designers, and craftspeople around the world. Whether creating monumental sculptures, delicate jewelry, or bespoke decorative items, the precision and versatility of the plasma cutter enable creators to push the boundaries of their craft and bring their most ambitious visions to life.

Repair and Maintenance Work

Plasma cutting, a technique renowned for its precision and efficiency, has significantly transformed the landscape of repair and maintenance work across various industries. The ability to cleanly and accurately cut through different types of metal with a plasma cutter has made it an indispensable tool in the arsenal of repair professionals and maintenance technicians. This powerful technology offers unparalleled advantages when it comes to restoring, repairing, and maintaining machinery, equipment, and structures, ensuring that they continue to operate at optimal efficiency and safety.

The versatility of plasma cutting shines in the realm of repair work, where the tasks can range from simple cuts to complex shapes and sizes needed for specific parts. Unlike traditional cutting methods, plasma cutting does not rely on physical force and therefore minimizes the distortion of the metal, resulting in cleaner, more precise cuts. This precision is critical when working on machinery or parts where every millimeter can affect functionality. For instance, in automotive repair, plasma cutters are used to remove and replace damaged sections of metal without compromising the integrity of the surrounding areas. This precision ensures that repairs are not only effective but also aesthetically pleasing, which is especially important in restoration projects.

Maintenance work also benefits from the efficiency of plasma cutting. In industries such as manufacturing, construction, and transportation, maintaining equipment and infrastructure is crucial to uninterrupted operation. Plasma cutters facilitate the quick removal of worn or damaged components and the fabrication of new parts on-site, significantly reducing downtime. The speed of plasma cutting, combined with its capability to cut through various metals, including steel, stainless steel, aluminum, and copper, makes it an invaluable tool in emergency repairs and routine maintenance checks.

Plasma cutting technology has been particularly transformative in sectors that require frequent updates or modifications to existing structures, such as the construction industry. It enables workers to make rapid adjustments to metal components, ensuring that projects can proceed without significant delays. Furthermore, the portability of many plasma cutting systems means that repairs and maintenance can be conducted in situ, eliminating the need to transport heavy or bulky items to a separate location for servicing.

Another application of plasma cutting in repair and maintenance is in the field of infrastructure, where it is used to cut through thick steel beams, plates, and other components. Whether it's repairing bridges, buildings, or public utilities, the ability to quickly and accurately cut through heavy-duty materials is essential for the restoration and upkeep of critical structures. The precision of plasma cutting not only speeds up the repair process but also contributes to the longevity and durability of the repairs.

Safety and environmental considerations are also paramount in repair and maintenance work, and plasma cutting addresses these concerns as well. The process produces less waste and fewer toxic fumes compared to traditional cutting methods, making it a more environmentally friendly option. Additionally, the reduced heat affected zone minimizes the risk of altering the properties of the metal being cut, ensuring that repairs maintain the strength and integrity of the original material.

In conclusion, the role of plasma cutting in repair and maintenance work is both transformative and expanding. Its ability to provide precise, efficient, and versatile cutting solutions has made it a key technology in maintaining the functionality and longevity of machinery, equipment, and infrastructure across a wide range of industries. As plasma cutting technology continues to evolve, its applications in repair and maintenance are set to become even more integral, further solidifying its status as an essential tool for professionals in these critical fields.

Chapter: 8 Health and Safety Considerations

Ventilation and Fume Extraction

The process of plasma cutting, while efficient and precise, generates fumes and gases that can be hazardous to health if not properly managed. The significance of ventilation and fume extraction in the context of plasma cutting cannot be overstated, as these measures are crucial to maintaining a safe working environment. Effective management of these byproducts is a multifaceted endeavor, requiring an understanding of the types of fumes generated, the potential health risks, and the systems and practices needed to mitigate these dangers.

Plasma cutting, by its nature, involves the melting of metal with a high-velocity jet of ionized gas. This action creates metal fumes and gases, including ozone, nitrogen oxides, and particulates from the material being cut, which can vary depending on the type of metal. For example, cutting through galvanized steel releases zinc fumes, while cutting stainless steel produces hexavalent chromium, a known carcinogen. The health risks associated with exposure to these substances range from short-term effects such as irritation of the eyes, nose, and throat, to long-term respiratory issues and increased risk of lung cancer.

Given these risks, the implementation of effective ventilation and fume extraction systems is critical. The goal of these systems is to capture fumes at the source and remove them from the breathing zone of the operator and others in the vicinity, thereby minimizing exposure. There are several strategies and equipment options for achieving this, each suited to different working environments and scales of operation.

At the most basic level, natural ventilation, achieved through open doors and windows, can provide a degree of fume management. However, this is typically insufficient for all but the most minor plasma cutting tasks. More commonly, mechanical ventilation systems are employed. These systems use fans and ducts to draw fresh air into the workspace and exhaust contaminated air outside. The effectiveness of mechanical ventilation depends on the design of the system, including the placement of intake and exhaust vents to ensure a flow of air that captures fumes efficiently.

For operations that require a higher level of control over fume extraction, localized exhaust ventilation (LEV) systems are a preferred option. LEV systems, such as fume extraction arms or downdraft tables, are designed to capture fumes at or near the point of generation. Fume extraction arms can be positioned close to the cutting area to capture fumes directly at the source, while downdraft tables draw fumes downward, away from the operator's breathing zone. These systems are particularly effective in enclosed or indoor environments where general ventilation may not be sufficient.

In addition to these systems, the use of personal protective equipment (PPE), such as respirators, may be necessary in situations where ventilation and fume extraction cannot adequately control exposure to harmful substances. The selection of appropriate respirators should be based on a risk assessment that considers the types of fumes generated and their concentrations.

Regular maintenance and testing of ventilation and fume extraction equipment are essential to ensure their continued effectiveness. This includes inspecting ducts and filters for blockages and wear, as well as monitoring airflow to confirm that it meets required standards. Furthermore, educating workers about the risks associated with plasma cutting fumes and the correct use of ventilation and protective equipment is vital to fostering a culture of safety.

In conclusion, the strategic implementation of ventilation and fume extraction measures is a cornerstone of safe plasma cutting practices. By effectively managing the hazardous byproducts of the cutting process, operators can protect their health and ensure a safer working environment. This underscores the importance of integrating comprehensive fume management practices into the broader framework of health and safety considerations in plasma cutting operations.

Noise and Ear Protection

The operation of a plasma cutter, while efficient and precise in cutting through various types of metal, generates significant noise levels. This aspect of plasma cutting is often underestimated but is crucial to consider for the health and safety of operators and bystanders. Prolonged exposure to high noise levels can lead to a range of auditory issues, including temporary or permanent hearing loss, tinnitus (a ringing in the ears), and an overall reduction in the quality of life. Thus, understanding the importance of noise and implementing effective ear protection strategies become fundamental to maintaining a safe working environment.

Plasma cutters produce noise through the high-velocity jet of plasma and the interaction between the plasma and the metal being cut. The intensity of this noise can vary depending on several factors, including the type of plasma cutter, the material being cut, the thickness of the material, and the cutting speed. Typically, noise levels can range from 90 to 120 decibels (dB), well above the threshold where prolonged exposure can cause hearing damage, which is 85 dB for an 8-hour exposure according to the National Institute for Occupational Safety and Health (NIOSH).

To mitigate the risks associated with noise exposure, a comprehensive approach to ear protection is necessary. The first line of defense involves the use of personal protective equipment

(PPE), specifically earmuffs or earplugs designed to reduce noise exposure to safe levels. Earmuffs provide a seal around the entire ear and are effective in reducing the volume of ambient noise. They are user-friendly and can be easily put on and taken off, making them convenient for workers who move in and out of high-noise areas. Earplugs, on the other hand, fit directly in the ear canal and can be more effective for individuals who find earmuffs uncomfortable, especially in hot environments or during extended use. For maximum protection, especially in environments where noise levels exceed 105 dB, the simultaneous use of both earmuffs and earplugs might be necessary.

Beyond personal protective equipment, engineering controls play a crucial role in reducing noise at the source. Enclosing the plasma cutter within a sound-dampening enclosure can significantly reduce the noise emitted into the surrounding environment. Additionally, manufacturers are continuously working to design quieter plasma cutting machines through advancements in technology and machine design.

Administrative controls also contribute to a comprehensive noise reduction strategy. Implementing a rotation schedule to limit the duration of exposure to high noise levels, maintaining plasma cutting equipment to ensure it operates efficiently, and designing the workspace to minimize sound reflection and amplification can all help reduce the risk of hearing damage.

Educating and training operators on the risks associated with noise exposure and the proper use of ear protection is essential. This includes understanding the correct way to use and maintain PPE, recognizing the signs of hearing damage, and knowing when and where ear protection should be worn.

In conclusion, while plasma cutting is a valuable tool in metal fabrication, the noise generated by its operation poses a significant risk to auditory health. Through the diligent use of ear protection, coupled with engineering and administrative controls, workers can safeguard their hearing. By prioritizing noise reduction and ear protection, the industry can ensure that plasma cutting remains not only an effective but also a safe method for cutting metal.

Eye Protection and UV Radiation

The process of plasma cutting, while efficient and precise, generates intense light, ultraviolet (UV) radiation, and infrared radiation. These emissions pose significant risks to the eyes, making eye protection a paramount concern for anyone using a plasma cutter. Understanding the nature of these risks and the protective measures required is essential for maintaining ocular health and ensuring a safe working environment.

UV radiation, invisible to the human eye, is one of the most hazardous byproducts of plasma cutting. Exposure to UV radiation can lead to photokeratitis, often referred to as "welder's flash" or "arc eye," a painful condition resembling a sunburn of the cornea. Symptoms include redness, a sensation of having sand in the eyes, excessive tearing, and sensitivity to light. In severe cases, prolonged exposure to UV radiation without adequate protection can lead to more serious eye injuries, including cataracts and permanent damage to the retina, which can impair vision.

To safeguard against these dangers, the use of an auto-darkening welding helmet equipped with a filter that meets the specific safety standards for plasma cutting is crucial. These helmets automatically adjust the darkness of the filter in response to the intensity of the light emitted, providing optimal protection against both the brightness and the UV radiation generated during the cutting process. The lens shade should be carefully selected based on the amperage of the plasma cutter, as higher amperages produce more intense light and require a darker shade to ensure adequate protection.

In addition to a welding helmet, wearing safety glasses with UV protection under the helmet offers an additional layer of defense, particularly against peripheral UV exposure or in the event the helmet is lifted. These glasses should comply with national safety standards for UV protection and ideally feature side shields to protect against UV exposure from all angles.

Beyond personal protective equipment, controlling the environment in which plasma cutting is performed plays a critical role in mitigating exposure to UV radiation. Enclosures or screens can be used to shield bystanders or other workers in the vicinity from the direct and reflected UV radiation. These barriers are especially important in workshops or facilities where multiple people may be present, as UV radiation can reflect off surfaces and pose a risk to those not directly involved in the cutting process.

Education and awareness are also fundamental components of a comprehensive eye protection strategy. Operators should be fully informed about the risks associated with UV radiation and the necessary precautions to take. This includes understanding the importance of regularly inspecting and maintaining protective gear to ensure it provides the intended level of protection.

In conclusion, protecting the eyes from the intense light and UV radiation produced by plasma cutting is an essential aspect of health and safety in metalworking. By employing the correct protective gear, such as auto-darkening welding helmets and UV-protective safety glasses, and taking measures to control exposure in the workplace, operators can significantly reduce the risk of eye injuries. This commitment to eye safety not only preserves the well-being of the individual but also enhances the overall safety culture within the metalworking environment, allowing for the safe and effective use of plasma cutting technology.

Proper Handling and Ergonomics

Proper handling and ergonomics are critical components of the health and safety considerations when using a plasma cutter. These considerations are designed to minimize the risk of injury and improve efficiency during the cutting process. Plasma cutting, while an exceptionally effective method for slicing through metal materials, requires the operator to manage equipment that is both heavy and intricate. This necessitates a focus on the ergonomic use of tools to prevent physical strain and ensure operational safety.

The design of plasma cutting equipment often integrates features aimed at reducing user fatigue and strain. For instance, modern plasma cutters and torches are constructed to be lightweight, with handles that conform to the natural grip of the hand. Despite these advancements, the duration of use, the repetitive nature of tasks, and the positioning of the body during cutting can lead to musculoskeletal disorders, particularly in the hands, wrists, elbows, shoulders, and back.

To mitigate these risks, it is essential to adopt a posture that keeps the body aligned and balanced while cutting. Operators should maintain a stable stance, with feet shoulder-width apart, to provide a solid foundation. This stance helps in managing the cutter's recoil and in maneuvering the torch smoothly over the workpiece without unnecessary strain.

When engaging in plasma cutting, it is imperative to adjust the height of the workbench to align with the operator's natural working posture. The workpiece should be positioned so that the operator does not have to bend forward excessively, reach overhead, or twist the torso and spine to view the cutting path clearly. Such adjustments reduce stress on the back and neck, lowering the risk of chronic injuries.

The technique of handling the plasma torch also plays a crucial role in minimizing ergonomic strain. Operators should use a relaxed grip, avoiding clenching or squeezing the torch too tightly, which can lead to hand and forearm fatigue. Additionally, using both hands can help distribute the weight of the torch more evenly, reducing the strain on any single arm.

Frequent breaks are another vital aspect of maintaining good ergonomic practices. Short breaks allow muscles to relax, especially after performing repetitive tasks or maintaining the same posture for extended periods. During these breaks, stretching exercises targeting the back, shoulders, and arms can help relieve tension and promote blood circulation.

Adjustable workstations that can be tailored to the task at hand and the individual's physical requirements are highly beneficial. Equipment such as adjustable tables, ergonomic floor mats that reduce the fatigue from standing, and properly adjusted lighting to reduce eye strain, all contribute to a healthier work environment.

Lastly, continuous education and training on ergonomic practices in the workplace can significantly enhance the awareness and adoption of proper handling techniques. Operators should be trained not only in the operational aspects of plasma cutting but also in recognizing the early signs of ergonomic strain and the best practices to mitigate them.

Incorporating ergonomic principles into the routine of plasma cutting operations is not just about preventing injuries; it's also about enhancing productivity and ensuring the longevity of a worker's career in metal fabrication. By prioritizing proper handling and ergonomics, workshops can maintain a safer environment that promotes the well-being of their operators and the efficiency of their operations.

Chapter: 9 Comparing Plasma Cutting with Other Cutting Methods

Plasma vs. Oxy-Fuel Cutting

The choice between plasma and oxy-fuel cutting is pivotal in metalworking, shaping not just the workflow but also the quality and efficiency of the outcomes. These two cutting technologies, while aiming to accomplish the same task—slicing through metal—do so through markedly different processes and have distinct advantages and challenges.

Plasma cutting, a technique that employs a jet of ionized gas at temperatures exceeding 20,000°C to melt and expel material from the cut, shines in its versatility and speed. This method is effective on a wide range of conductive metals, including stainless steel, aluminum, and copper, offering significant flexibility. Its ability to perform precise cuts at remarkable speeds is unparalleled, especially when cutting thin and medium-thickness metals. Furthermore, plasma cutters are capable of piercing metal almost instantaneously and can be easily adapted for automation, enhancing both precision and productivity.

On the other hand, oxy-fuel cutting relies on a chemical reaction between oxygen and the metal being cut to generate heat, effectively burning through the metal. This traditional method excels with specific materials, particularly carbon steel, and can handle thicker materials more efficiently than plasma cutting. The equipment for oxy-fuel cutting is typically less expensive and simpler, making it accessible for a wide range of applications, especially in environments where electricity is not readily available.

When comparing the two methods, several key differences emerge:

1. **Material Compatibility**: Plasma cutting's ability to cut through any conductive material gives it a broader application range. Oxy-fuel, though limited to primarily carbon steel, cuts thicker materials more economically.

2. **Cutting Speed**: Plasma cutting is generally faster, especially on materials up to 2 inches thick. For very thick materials, oxy-fuel might have the edge due to its slower, but steady cutting speed.

3. **Precision and Cut Quality**: Plasma cutting tends to produce less slag and can achieve more precise cuts with a narrower kerf.

The quality of cut edges is also generally better, reducing the need for secondary finishing operations.

4. **Portability and Setup**: Plasma cutting equipment is more complex and requires a power source and compressed air, whereas oxy-fuel equipment is simpler and highly portable, ideal for field work.

5. **Operating Costs**: Plasma cutters often have higher initial costs and require consumables like electrodes and nozzles. However, the speed and efficiency of plasma cutting can lead to lower overall operating costs in many applications. Oxy-fuel's consumables are typically less expensive, but the process is slower and less efficient on thinner materials.

6. **Safety**: Both methods require stringent safety precautions. Plasma cutting produces ultraviolet light and requires eye protection, while oxy-fuel cutting involves handling and storing highly flammable gases.

In choosing between plasma and oxy-fuel cutting, the decision hinges on the specific requirements of the project, including the type and thickness of the material, the desired precision, the available equipment, and the working environment. Plasma cutting offers a modern, versatile, and fast option for many applications, particularly where precision and efficiency are paramount. Oxy-fuel cutting, while more traditional, remains a cost-effective choice for specific applications, particularly with thick carbon steel where its slower pace and simplicity are not detrimental.

Ultimately, the choice reflects a balance between these factors, underscoring the importance of understanding the capabilities and limitations of each cutting method to select the most appropriate for the task at hand. This knowledge enables metalworkers to optimize their cutting processes, ensuring quality, efficiency, and safety in their operations.

Plasma vs. Laser Cutting

When comparing plasma cutting with laser cutting, it's crucial to delve into the unique characteristics, benefits, and limitations of each method to understand their respective roles in metal fabrication. This comparison not only highlights the versatility and precision these technologies offer but also helps users decide which method best suits their specific project needs.

Plasma cutting operates by sending an electric arc through a gas that is passing through a constricted opening. The gas can be air, nitrogen, argon, or a mixture, which is heated to an extremely high temperature and ionized so that it becomes plasma. This plasma, sufficiently hot to melt the metal being cut and moving at a high velocity, blows the molten metal away, thereby cutting through the workpiece. Plasma cutting is renowned for its efficiency in cutting thick materials quickly and cost-effectively.

On the other hand, laser cutting employs a high-powered laser beam focused through optics to cut materials. The laser beam has a high energy density, making it possible to melt, burn, or vaporize the material. Laser cutting is celebrated for its precision, capability to cut small and intricate shapes, and the high quality of the cut edges, which typically require minimal to no finishing work.

Efficiency and Material Thickness

Plasma cutting excels in cutting thicker materials, especially where the edge quality is not the primary concern. It is faster than laser cutting when working with metal thicknesses over 1/2 inch. In contrast, laser cutting provides superior precision but its efficiency decreases as the material thickness increases. Laser cutters, particularly CO_2 lasers, can cut thicker materials but at a slower pace and with higher operating costs compared to plasma cutting.

Cut Quality and Precision

Laser cutting stands out for its high precision, capable of producing intricate cuts and fine details with a smaller kerf (cut width) compared to plasma cutting. The laser's ability to focus to a small point enables it to achieve highly detailed cuts, making it ideal for applications requiring intricate designs and tight tolerances. Plasma cutting, while improved with modern technology, still has a wider kerf and may produce a somewhat rougher edge, which might require additional finishing for some applications.

Material Compatibility

Plasma cutting is predominantly used for conducting materials such as steel, stainless steel, aluminum, brass, and copper. It's versatile and effective across a broad range of thicknesses but is generally not used for non-metallic materials. Laser cutting, however, is more versatile in terms of material compatibility. It can cut through metals, plastics, wood, glass, and even rubber, making it a more flexible choice for projects that involve multiple types of materials.

Operational Cost and Maintenance

The operational costs of plasma cutting are typically lower than those of laser cutting. Plasma cutting equipment is generally less expensive to purchase and maintain, and it consumes less energy compared to a high-powered laser cutter. However, the cost advantage of plasma cutting must be balanced against the specific requirements of the project, such as the need for precision and the types of materials to be cut.

Safety and Environmental Considerations

Both methods require adherence to safety protocols to protect operators from potential hazards. Plasma cutting generates bright ultraviolet light, necessitating protective eyewear, and produces fumes that must be ventilated. Laser cutting also requires eye protection to guard against harmful laser light, and depending on the material being cut, it may produce hazardous fumes or dust that require adequate extraction.

In summary, the choice between plasma and laser cutting hinges on the specific requirements of the project, including the material type and thickness, desired cut quality and precision, budget constraints, and the available equipment. Plasma cutting offers speed and cost-efficiency for thicker materials, making it ideal for heavy-duty industrial applications. Laser cutting, with its high precision and versatility, excels in detailed and intricate designs across a variety of materials, suitable for both industrial and artistic applications. Understanding these differences enables users to select the most appropriate cutting method, ensuring optimal outcomes for their projects.

Advantages and Limitations

Plasma cutting stands out in the pantheon of metal cutting methods for its unique combination of speed, efficiency, and versatility. This technique, which uses a jet of superheated, electrically ionized gas to slice through metal, offers several advantages over traditional cutting methods like oxy-fuel cutting and more modern techniques like laser cutting. However, like any technology, it also comes with its limitations, making it important to understand both sides to fully appreciate how it fits into the broader context of metalworking.

One of the primary advantages of plasma cutting is its versatility. Unlike oxy-fuel cutting, which is limited to cutting only ferrous metals due to its reliance on oxidizing the metal to cut through it, plasma cutting can slice through virtually any type of conductive metal. This includes steel, stainless steel, aluminum, brass, and copper, making it a highly versatile tool for a wide range of applications. This capability is particularly valuable in shops that work with a variety of metals, providing a single tool solution for multiple cutting needs.

Speed is another significant advantage of plasma cutting. In most cases, plasma cutting is much faster than oxy-fuel cutting, especially for thinner materials. This increase in speed can dramatically reduce project times, improving efficiency and productivity. When compared to laser cutting, plasma cutting also holds its own, especially for medium-thickness materials, offering competitive speeds without the high operating costs associated with laser cutters.

The initial setup and operating costs of plasma cutting are generally lower than those of laser cutting systems. While high-end, industrial-grade plasma cutters can represent a significant investment, entry-level systems are accessible for smaller shops and hobbyists, offering a lower barrier to entry than laser cutting systems. Additionally, the operating costs, including power and consumables, tend to be lower for plasma cutting, making it a cost-effective option for many metal cutting applications.

However, plasma cutting is not without its limitations. When it comes to precision and cut quality, laser cutting often holds the upper hand, especially for very fine details and extremely thin materials. The heat-affected zone (HAZ) is larger in plasma cutting, which can lead to more thermal distortion compared to laser cutting. This aspect can be a critical consideration for applications requiring high precision or for materials sensitive to heat.

Another limitation of plasma cutting is the noise and production of fumes and particulates. The process is louder than laser cutting and generates more smoke and particulates, necessitating good ventilation systems to maintain a safe and comfortable working environment. While modern plasma systems have made strides in reducing these issues, they still remain a consideration, especially in enclosed or poorly ventilated spaces.

Moreover, while plasma cutting is capable of cutting through thicker materials than many laser cutters, it does so at the cost of edge quality and precision. For very thick materials, the kerf width (the width of the material removed by the cutting process) can be wider, and the cut edges might require additional finishing operations.

In conclusion, plasma cutting offers a powerful, versatile, and cost-effective solution for cutting conductive metals, striking a compelling balance between speed, versatility, and operational costs. Its advantages make it a go-to choice for a wide range of industrial, commercial, and artistic applications. However, its limitations in terms of cut quality and precision, as well as environmental considerations, mean that it may not always be the best choice for every application. Understanding these strengths and weaknesses is crucial for anyone looking to integrate plasma cutting into their metalworking practices, ensuring they can make the most of this dynamic cutting method while navigating its challenges.

Conclusion

In wrapping up the exploration of plasma cutting, it's evident that this technology has significantly revolutionized metalworking, offering a blend of speed, precision, and versatility unmatched by many traditional cutting methods. The journey from understanding the basic principles that underpin plasma cutting to mastering its advanced applications showcases its potential to not only enhance productivity but also to expand creative possibilities.

The core appeal of plasma cutting lies in its ability to cut through various types of conductive metals with remarkable efficiency. This capability opens up a wide range of applications, from industrial manufacturing and automotive repair to artistic metal sculpture. The technology's accessibility further broadens its appeal, with equipment options available that cater to high-end industrial users as well as hobbyists and small workshops. This inclusivity ensures that the benefits of plasma cutting can be leveraged across a broad spectrum of users.

However, the successful adoption of plasma cutting is not without its challenges. The process requires a thorough understanding of the equipment, as well as a commitment to maintaining strict safety standards. Protective gear and proper ventilation are essential to safeguard against the potential hazards associated with the process, such as exposure to intense light, harmful fumes, and the risk of burns. Moreover, achieving optimal results with plasma cutting demands a nuanced understanding of factors like material thickness, cutting speed, and the correct setup of the equipment.

Despite these challenges, the advantages of plasma cutting often outweigh the limitations. The technology offers a cost-effective alternative to more expensive cutting methods like laser cutting, without compromising significantly on speed or quality for a wide range of applications. Its efficiency and versatility make it an indispensable tool in the metalworking industry, capable of delivering precise cuts and intricate details that meet the demands of both functional and artistic projects.

Furthermore, the ongoing advancements in plasma cutting technology continue to expand its capabilities and reduce its limitations. Innovations in cutter design, consumables, and software are making plasma cutting more precise, less resource-intensive, and easier to use. These developments promise to further solidify plasma cutting's position as a key technology in metalworking, capable of meeting the evolving needs of industries and artists alike.

In conclusion, plasma cutting represents a dynamic and powerful method for cutting conductive metals, offering a unique combination of speed, precision, and versatility. Its application spans a wide array of industries and disciplines, underscoring its importance in modern metalworking. While it comes with its own set of challenges and limitations, the continued evolution of plasma cutting technology suggests a bright future, with even greater potential for innovation and application. For those willing to invest the time to learn and master this technology, plasma cutting offers the opportunity to transform metal into works of precision and artistry, making it an invaluable tool in the arsenal of modern metalworkers.

www.ingramcontent.com/pod-product-compliance
Lightning Source LLC
Chambersburg PA
CBHW050305230526
45471CB00005B/2036